Birds of the Lake Erie Region

Birds of the Lake Erie Region

Carolyn V. Platt

photography by Gary Meszaros

THE KENT STATE
UNIVERSITY PRESS
Kent, Ohio, & London

Library of Congress Catalog Card Number 00-062022
ISBN 0-87338-690-6
Manufactured in China

05 04 03 02 01 5 4 3 2 1

Frontispiece: Canvasbacks stage in rafts around Lake Erie.

Library of Congress Cataloging-in-Publication Data
Platt, Carolyn V., 1943–
Birds of the Lake Erie region / Carolyn V. Platt ;
photography by Gary Meszaros.
p. cm.
isbn 0-87338-690-6 (alk. paper)
1. Birds—Erie, Lake, Region. I. Title.
QL685.5.06 P53 2001
598'.09771'2—dc21 00-062022

British Library Cataloging-in-Publication data are available.

Contents

Foreword

The vivid photos and text of *Birds of the Lake Erie Region* bring back my own glowing memories from trips to the area's extraordinary birding spots. Although a Californian for most of my life, I've always loved watching birds around Lake Erie and went there often with Sue Tackett, Charlotte Mathena, Larry Rosche, and other companions when I lived for a while in southwestern Ohio. I'll certainly make many more return treks to this remarkable region.

My first birding trip to Lake Erie was on December 11, 1980, when we drove up to Presque Isle State Park from Pittsburgh. At Presque Isle, Pennsylvania's finest birding site, we met a young lad named Jerry McWilliams. Over the years Jerry, who is coauthor of *The Birds of Pennsylvania*, has made many important discoveries there.

On this December day two decades ago, birding conditions were hardly ideal. Snow had fallen for much of the day and was still coming down heavily when we arrived in the afternoon; a gale howled out of the northwest and blew the snow almost horizontally. All was quiet except for wind and the sounds of a few black-capped chickadees and yellow-rumped warblers. At the edge of a field, we found American tree sparrows and in the woods a roosting great horned owl. Offshore we spotted a female surf scoter among many other ducks and witnessed the glorious sight of 500 tundra swans. Braving the wind and snow, we hiked out to the point and were rewarded with a lone snowy owl, a dark immature female.

I came back to Lake Erie the following May, this time to Point Pelee on the north shore, among Canada's most celebrated birding sites. I was co-leading a tour with one of Ontario's best birders, Bob Curry. From such famous names as Roger Tory Peterson, in my teens I'd read about the magnificent spring migration at Point Pelee so I knew that weather is a primary factor in the number of migrants appearing there. On "bluebird" days, a cold wind out of the north can keep most birds south of the lake and leave Pelee almost empty, but on my first trip I was not to be disappointed.

There were migrants everywhere! I had never soon so many flycatchers, vireos, thrushes, warblers, scarlet tanagers, and Baltimore orioles. The trees had not yet leafed out, and the birds were quite visible, giving us great chances to compare similar species. These included spotted Catharus thrushes, the eastern species of Empidonax flycatchers, warbling and Philadelphia vireos, and a northern and Louisiana waterthrush sharing the same pool of water. For the first time, I realized just how pink the legs of a Louisiana were. Another bird I won't soon forget was the Henslow's sparrow skulking in the grass just a few feet away.

We observed over thirty species of warbler, and most of the males were in full song. While other locations, such as High Island in eastern Texas and the Dry Tortugas off Key West, Florida, are well-known hot spots for warblers, the mix of species in the Lake Erie region and on the southern edge of Lake Michigan is larger. At High Island you can see trans-gulf migrants, but you'll miss both West Indian–Florida migrants and the circum-gulf migrants that travel up through south Texas.

At Lake Erie all routes converge. I've personally tallied thirty-three species of warbler in a day at Point Pelee, and my companion Bob Brackett, from Ottawa, has seen thirty-four. It's certainly no accident that I've come back to this great site every year since and did most of my field research for *Warblers* (Houghton-Mifflin, 1997) around Lake Erie.

Though Pelee is the most renowned place for spring migration, other hot spots, such as Long Point, Ontario; Presque Isle State Park, Pennsylvania; and Crane Creek State Park in Ohio are also superior places to bird. The boardwalk at Magee Marsh Wildlife Area next to Crane Creek can rival the numbers of migrants seen at Pelee. Many prefer this boardwalk, since views of birds are usually better and not so dependent on weather conditions—you can consistently find good numbers on most days. Workers at the bird observatory at Ontario's Long Point have gathered massive scientific data over a long period at that site and have noted many great rarities, such as black-capped vireo, varied bunting, and hooded oriole.

The spring migration keeps on well into June for some species, but in July interest transfers to shorebirds' movements southward. Depending on Erie's water levels, this can be exciting too. Most are species birders expect, such as pectoral sandpiper, Earth's champion long-distance traveler. However, rarities like snowy plover, curlew sandpiper, and ruff sometimes turn up as well.

Pointe Mouillee, south of Detroit in Michigan, is probably the best place to see shorebirds around Lake Erie now. Metzger Marsh Wildlife Area east of Toledo, Ohio, was once superb as well. Here hundreds of long-billed dowitchers, scarce elsewhere in the Midwest, were discovered in the early 1990s. I counted nearly 500 on several occasions. Unfortunately, Ducks Unlimited, backed by government agencies, rebuilt a dike across the marsh to boost use by waterfowl. Phragmites grass soon took over the area, which is now greatly degraded and useless for shorebirds. Even those responsible for this ecological disaster don't yet know how to repair the damage.

As summer turns to fall, passerines move south in great numbers. Small flocks of snow buntings arrive around the lake's edge by the end of October, and in some years snowy owls and flocks of redpolls also appear. In November thousands of red-breasted mergansers and Bonaparte's gulls stage along the shore. Birders can usually find one or more little gulls and the odd black-legged kittiwake by carefully searching through the Bonaparte's. They may also see something rarer, like a black-headed or Ross's gull. As the lake ices in, most of the small gulls leave Lake Erie and migrate to the mid-Atlantic coast; larger species like herring, lesser black-backed, and the white-winged gulls (glaucous, Iceland, and Thayer's) replace them.

Niagara River Gorge is the most exhilarating gull-watching spot in North America, with thousands of gulls massing just above and below the falls. Observers have recorded nineteen species there. The staff at *Birders Journal* hosted a gull conference at Niagara this past November—over 200 gull enthusiasts attended from as far away as the Netherlands and Sweden. Who says nobody bothers to look at gulls?

There is, of course, much more to say about the bird life of Lake Erie, especially the tremendous passage of hawks and other raptors at points along the north shore in fall. In remnant patches of Ontario's Carolinian forest, several songbirds reach the northern edge of their breeding ranges; both breeding and winter ranges have changed a great deal around the lake in the past two centuries, an absorbing story of its own.

But all of these subjects and more are covered in detail in this book. Carolyn V. Platt, a longtime resident who has written extensively about the natural history of Lake Erie's environs supplies lucid text. The exceptional photos are by Gary Meszaros, one of North America's finest nature photographers, who has spent much of his life photographing the region's rich plant and animal life. This winning combination makes *Birds of the Lake Erie Region* a fine resource for anyone interested in the area's fascinating bird life.

Jon L. Dunn
Bishop, California

Acknowledgments

First, we want to thank our spouses, Eric Hoddersen and Jane Meszaros, who were so supportive in this endeavor, and Gary's two daughters, Amy and Carrie.

Special recognition goes to Allen Chartier of the Holiday Beach Migration Observatory, who found time from his many obligations to review the manuscript and answer questions. His thoughts and good ideas greatly enhanced our book. Allen, we couldn't have done this without you!

Gratitude is also due Fred Urie of Windsor, Ontario, who answered innumerable questions about birds on the north side of the lake. Fred also reviewed the manuscript for accuracy. Rob Harlan's help was invaluable as well: His years as editor of our state bird publication, *The Ohio Cardinal,* helped us ferret out persistent little inaccuracies. John Pogacnik and Alan Wormington read portions of the manuscript and willingly shared their thoughts and information.

Many other people contributed directly or indirectly by calling about rare birds, sending newspaper clippings, offering information, or answering questions. They include Ian Adams, Alice and Tom Faren, Vic Fazio, Joyce and Anders Fjeldstad, Tom LePage, Jerry McWilliams, Ed Pierce, and Bob Segedi.

Gary also salutes his longtime friends in the field, Andy McClure and Edward Stroh, nest finder extraordinaire. Carolyn thanks Joy Kizer and all the other librarians and trip leaders who have fed her lifelong appetite for information; in addition, she gratefully recognizes Cuyahoga Community College for supporting projects rather unusual for a teacher of English.

Introduction

For the animal shall not be measured by man. In a world older and more complete than ours they move finished and complete, gifted with extensions of the senses we have lost or never attained, living by voices we shall never hear. They are not brethren, they are not underlings; they are other nations, caught with ourselves in the net of life and time, fellow prisoners of the splendour and travail of the earth.

—Henry Beston, *The Outermost House*

Gary Meszaros and I have collaborated on this volume, *Birds of the Lake Erie Region*, firstly because we hope to give readers a glimpse of the "splendour and travail," in Henry Beston's words, of birds in the area that lies immediately around our great lake. Gary's photographs, taken over decades of rising before dawn and spending myriad hours in the field in all weathers, effectively present their visual splendor. The chapters' text introduces the tale—sometimes almost beyond belief—of these remarkable creatures' migration waves, their summer nesting populations and behaviors, and their strategies for surviving cold and dearth in the hunger season. An appendix lists and briefly describes premier birdwatching locations around the lake.

We also want those who open our book to understand the unity of the region and the major effects of the lake itself on both resident birds and migrants. Lake Erie moderates weather, allowing the eastern deciduous forest to extend farther north than it otherwise would. That forest zone surrounds the lake, creating a relatively consistent natural setting, though greatly altered by human beings. These factors strongly affect resident species' distribution. The lake also provides a highway and staging ground for some migrating birds while posing a formidable barrier to others.

Though it makes very good sense to treat Lake Erie's region as a whole when thinking about bird life, this is seldom done in any integrated way. Political boundaries often condition the ways we think about natural regions and our relation to them—what

Canadian would agree with northeastern Ohio's self-designation as the "North Coast"? I know that for this Ohioan, one of the most satisfying aspects of the project has been to focus on the natural setting of southern Ontario, as well as on those of neighboring states.

Point Pelee National Park and Long Point on the north shore of Lake Erie in Ontario are internationally recognized birding hot spots, two of the best sites in North America to observe spring and fall migrations. However, a further reason we produced this book is to show that glorious as the spring landbird migration may be at these spots, there is much more to birding in the Lake Erie region: Sensational concentrations of gulls gather on the Niagara River in autumn, and birders have only recently grasped the importance of fall raptor migration through the area. It is as good, if not better, than at more celebrated spots like Hawk Mountain in eastern Pennsylvania and Cape May, New Jersey. Waterfowl concentrations are spectacular here, especially in spring, and in certain years shorebirds can be abundant.

Bird observatories such as Ontario's Long Point and, more recently, Holiday Beach Migration Observatory, Ontario, Southeastern Michigan Raptor Research, and Ohio's Black Swamp Observatory have done fine work in collecting data on migration. Volunteers have gathered much of the data, it should be noted, and thus have added vital knowledge about birds' movements and the region's natural history.

Of particular concern to us are the problems of wetlands around Lake Erie today. Explosions of alien species like European starlings and carp, as well as natives such as ring-billed gulls, common grackles, Canada geese, and muskrats, have severely restricted the more specialized birds and have already eliminated a number of them as breeding species. Problems of invasive plants like phragmites, canary grass, and purple loosestrife must be addressed if we are to save our remaining cattail marshes. Many managers seem to lack the expertise or perhaps the resources to employ the active management practices needed.

We must stop judging the quality of our marshes by how many Canada geese are harvested each year. A good example: In Ohio's Mallard Club Marsh, one of the few locations on Erie's south shore that still hosts nesting bitterns and rails, phragmites (giant reed grass) is gradually overtaking cattail areas. These areas will probably disappear by the end of the decade. Wetland problems seem monumental in the increasingly urbanized and intensively farmed environment around the lake. On the brighter side, enhancements are now being made to many marshes, like those at Michigan's Pointe Mouillee and at Pipe Creek and Pickerel Creek in Ohio. Though the wetlands created are less than perfect, this is, nevertheless, a move in the right direction.

Birds of the Lake Erie Region is organized roughly to follow the year's seasons, beginning and ending with winter. "Birds and Ice" describes the great numbers of ducks and other waterbirds that converge to feed in Lake Erie's last ice-free areas during extreme winter weather. The second chapter, "Landbird Fallout" addresses spring migration, especially the incredible showers of passerines at Point Pelee, Ontario, Magee Marsh, in Ohio, and Presque Isle in Pennsylvania, in May. Geographical and historical perspectives inform "Area Nesters," which discusses how the lake and the Eastern Deciduous Forest Zone affect nesting distribution. This chapter also shows the powerful effects of human settlement and environmental exploitation on area bird populations over time.

"The Lake Erie Marshes," with an appreciation of these wetlands' diverse bird life, explores the marshes' original vastness and variety and traces their evolution and drastic shrinkage since settlement. Included are the cycle of marsh seasons and treatments of both migrant and resident birds, especially waterfowl. Fall migration is discussed in "The North Wind," including raptor and shorebird movements. (We borrowed the title from Holiday Beach Migration Observatory's newsletter of the same name.) Entranced by the spring spectacle, birders sometimes fail to appreciate quieter dramas of the autumn season. Finally, the last chapter, "Against the Odds," examines and celebrates survival tactics of the area's winter residents and turns the corner of the year to anticipate spring once again.

Birds of the Lake Erie Region is truly a collaborative work. Gary and I met in 1981 when we were both contributing to the Publications Department at the Cleveland Museum of Natural History. Our first joint ventures were articles and photographs for the museum's magazine, *The Explorer*. In 1984 we began submitting pieces to *Timeline*, the magazine of the Ohio Historical Society, and an essential resource for anyone interested in the state's history or natural history. We still work with the magazine's staff today. Several chapters of this volume are reworked from articles originally appearing in *Timeline*. Our first book-length collaboration, *Creatures of Change: An Album of Ohio Animals*, appeared in 1998, published by Kent State University Press.

The natural world and watching birds have fascinated both Gary and me since childhood, and we hope that this book demonstrates the depth of our involvements. My own strength is of the magpie variety: I love picking up shiny bits of information, bringing them back to the nest, and laying them out in patterns meaningful to me and—I hope—interesting to others. I'm definitely a "big picture" sort of person. Gary is a visual artist and, equally important, a rare field naturalist. I often rely on his ideas for inspiration, as well as on his birding skills for primary information and as a check on my own accuracy. We have consulted on all chapters, and especially on photo captions and the bird-finding appendix.

Gary took the photographs over a twenty-five-year period. In almost all cases, they were shot in the wild and around Lake Erie using available natural light. We have tried to emphasize how viewers might actually see these species. Unfortunately, we could not include photographs of all the birds found in the region (well over 300) in a book of this size. *Birds of the Lake Erie Region* is not meant to be a bird identification guide—many excellent ones already exist. Instead, we have highlighted species relevant in telling the story of Lake Erie and its rich bird life. We hope you like this book as much as we've enjoyed working with Lake Erie's birds, birders, and the Kent State editorial staff to put it together.

Birds of the Lake Erie Region

Birds and Ice

The ice was here, the ice was there,
The ice was all around:
It cracked and growled, and roared and howled,
Like noises in a swound!

—Samuel Taylor Coleridge,
The Rime of the Ancient Mariner

The time is midwinter in northern Ohio, and a pale sun hangs like an ice ball over Lake Erie's vast white surface. Nothing blocks the razor wind cutting quickly to the bone. During severe winters, temperatures may drop close to minus twenty degrees Fahrenheit (-29°C) and the wind-chill factor to minus fifty degrees or lower. Great piles and blocks of ice shimmer blue with cold. As the light fades, lake currents pull at the treacherous pack ice, which cracks like thunder to reveal a widening hole of frigid steel-blue water. Some of these holes are kept open by currents, others by shifting winds. When open water meets frigid air, clouds of smoke form; termed "sea smoke," they can sometimes span the lake's horizon.

Natural openings in the ice, as well as those created by warm water from shoreside power plants, mean survival for thousands of wintering bay ducks, mergansers, and gulls that mass within them. Many migratory waterfowl fly only as far south in winter as they must to find open water. The five Great Lakes, America's largest inland bodies of fresh water, form almost continuous shorelines from New York to Minnesota, from eastern to western Ontario, and provide a natural highway for migrating waterbirds. When winter weather is severe, as it was in the terrible seasons of 1976–77 and 1977–78, a million ducks may funnel

Canvasbacks stage in big rafts at key locations around Lake Erie. These may hold thirty thousand birds. Better management and a string of wetter-than-normal years on the plains have raised canvasback numbers recently.

(*Opposite*) This scene from the power plant at Avon, Ohio, shows how critical open water is to wintering birds. When temperatures plummet, the warm waters from the plants create openings that can fill with thousands of ducks and gulls.

Black ducks and mallards explode from Castalia, Ohio's artesian spring-fed pond. In an average winter, over 10,000 "puddle" ducks may gather here. During the cold winters of 1977–79, numbers doubled, with many birds perishing from starvation.

south through the St. Lawrence River and Lake Ontario in search of Lake Erie's last ice-free areas.

They are searching not for warmth but for food—the small fish that Lake Erie produces more abundantly than any other great lake, accessible only in the openings among the huge chunks of ice. The masses of gulls and ducks that crowd into these shrinking holes treat hardy observers to a dazzling display: common goldeneyes, scaups, canvasbacks, redheads, buffleheads, legions of red-breasted mergansers, common mergansers, and herring gulls to glory. Rarer Arctic birds that occasionally winter on Lake Erie are visible as well: glaucous and Iceland gulls and beautifully feathered oldsquaws (recently renamed long-tailed ducks), sea ducks that flash their lovely dark brown, pearl gray,

Ring-necked ducks begin to migrate after the ice leaves most wetlands. Observers can sometimes see flocks of over a thousand birds on inland lakes and reservoirs during March and April.

A first-winter glaucous gull pecks at a dead merganser that failed to survive the night. Large predatory gulls quickly dispose of sick and dying birds.

These bachelor redheads are vying for the attentions of an as-yet-unmated female. Most pairings are made before the ducks reach breeding grounds. Redheads are known for depositing their eggs in the nests of other ducks.

Myriad ring-billed gulls swarm at a river mouth. Our most common gull species, thousands can be viewed in an average winter. Ring-billeds' populations have swelled here during the past half-century, and they have successfully competed against common terns, contributing to the terns' decline.

and white plumage among the ranks of plainer ducks, continuously squawking and yodeling their loud three-part calls.

All these birds have a desperate struggle ahead. Since only small areas of open water remain, food dwindles, and competition becomes fierce. Gulls fight over scraps of fish filched from diving ducks. They also wait to finish off any old or weakened ducks that cannot survive the coming night. Larger gulls like great black-backed and glaucous gulls dominate the scene, grabbing fish away from smaller and younger birds. A few great blue herons that decided to tough out the winter can also be seen along the edge of the ice, their bodies hunched over, trying to conserve as much body heat as possible. Other birds may come out to the edge of the ice to hunt—bald eagles to search for dead fish and peregrine falcons to snatch an occasional duck or gull, a change from their usual diet of urban rock doves.

An adult double-crested cormorant must constantly keep ice from forming on its plumage. Juvenile birds usually lose many wing and tail feathers, which limits their ability to catch fish and in turn leads to high mortality among young wintering birds. The lake's cormorant population exploded during the 1990s.

Greater scaup land in an ice-free hole on Sandusky Bay, Ohio. They winter there in large numbers along with goldeneyes and lesser scaup.

All the birds are well adapted to face the period of black cold that approaches as the sun sets. Their feathers provide the principal means to regulate body temperature. Birds' plumage is densest in winter, and they can fluff it out to conserve body heat. When birds sleep, they often tuck their bills beneath feathers to lower heat loss. Breathing rate and metabolism drop, and fat layers both insulate the body and provide necessary energy. The structure of these birds' feet also cuts down on heat loss. Feet and toes have no fleshy muscles, as human extremities do, but are formed of tough tendons with few nerves and limited blood vessels. In the cold water, blood supply cuts off, and the temperature of feet and legs drops nearly to freezing without causing damage.

But adaptation goes only so far. Old birds, weak birds, and those that cannot catch enough food to keep body heat at safe levels, will die before sunrise. (During the winter of 1976–77, each morning revealed more bodies of ducks and gulls that had died during the night.) The weakest of the still-living will make a quick breakfast for ravenous gulls.

Dabbling or "puddle" ducks have also passed a rough night not far away at the spring-fed duck pond area at Castalia, Ohio, near Sandusky and at the few other inland bodies of water not yet iced in. Five artesian springs, whose temperatures stay between forty-six and fifty-one degrees Fahrenheit year-round, keep Castalia's water ice free, even on this cruelly frigid morning. Approximately 10,000 ducks—mallards, American black ducks, and American wigeons—may winter there. The spot is especially good for viewing black ducks, a species that has dwindled

Red-breasted mergansers' serrated bills are ideally suited to catch the plentiful supply of shiners in Lake Erie, and many males winter here. Most females shun the cold weather and sojourn farther south.

These herring gulls wear breeding or alternate plumage. Herring gulls are abundant winter visitors to the region, but as winter wanes many migrate back up the St. Lawrence River.

Pied-billed grebes commonly breed in the Lake Erie marshes. They are sometimes grounded when they mistake wet roads for bodies of water and are unable to launch themselves without pattering over the surface of water.

steadily since the mid-1950s. With the Tennessee River Valley, this part of the Lake Erie coast is the black ducks' main wintering spot in the United States interior; another important wintering area for them is Quebec's upper St. Lawrence region. Competition from the more aggressive mallard keeps their numbers on the downward spiral. During the past decade, hundreds of tundra swans have begun to winter at Lake Erie Metropark south of Detroit, Michigan. Thousands of migrant swans will join them in March, feeding in farm fields north of Chatham, Ontario, and around Wallaceburg, Ontario.

The miracle is that so many birds do survive the lethal winter weather. Luckily, such cold snaps are short lived. The weather soon moderates, and south winds begin to open the lake ice packs, easing the fierce competition. Strong southerly breezes may pile ice over twenty feet high at places like Long Point and Point Pelee on Erie's northern shore. With the first thaw, thousands of waterfowl begin to move, following the retreating ice on their age-old route to the pothole country of the northern interior with tundra swans in the vanguard. Birders can see rafts of canvasbacks in the Detroit River, off Long Point, Ontario, on Sandusky Bay in Ohio, and at further staging areas such as the Niagara River. Thousands of redheads, ring-neckeds, and scaups of both species also mass on Erie's bays and inlets. Mergansers gather in dramatic numbers as well. Huge rafts of common mergansers can be found most years just offshore of Monroe County, Michigan.

Other ducks that have wintered to the south begin to appear on daily tallies by mid- to late February and early March. Winter residents that a few weeks earlier were fighting for their lives have departed for parts unknown, and only small flocks of gulls and waterfowl still dot the bays and inlets. Buffleheads are among the last to leave, some waiting well into April to depart for northern breeding grounds.

For certain birds migration is quite long, with some tundra swans traveling as far as Siberia. Others fly to the Yukon Delta. Canvasbacks and redheads head for North Dakota's prairie potholes and the Canadian interior. Lake Erie's geography swells the numbers: The lake lies across two great flyways, the Atlantic and the Mississippi. Birds coming up from the Chesapeake Bay and Gulf states converge over Erie's western basin before continuing

(*Overleaf above*) Lesser scaup, called "bluebills" by hunters, are common here in winter. Thousands gather in the lake's western basin during March and April. The numbers apparently increase each year; burgeoning zebra mussels seem to provide an excellent food source.

(*Overleaf below*) Birders tally large numbers of buffleheads in autumn with wintering birds breaking up into smaller feeding groups. The ducks may tarry in the area until April before moving north.

Common mergansers line the edge of the ice. Masses of them appear just offshore in Monroe County, Michigan, most winters.

north. The larger gull species begin to make their slow retreat back across Lakes Erie and Ontario and up the St. Lawrence River.

Another winter is winding down on the Lake Erie shore, and the frenzied nesting season is at hand. Far to the north, this year's survivors are busy replacing last winter's casualties with a vigorous crop of young ones, which next year may be fighting to survive another winter among Lake Erie's blocks of steel-blue ice.

The first southerly breezes bring flocks of migrating tundra swans, with the main push coming in early March. Fields around Lake St. Clair Wildlife Area in Ontario are a major staging area. If the winter is mild, hundreds may stay in the region.

(*Opposite*) On frigid mornings, mallards try to conserve energy by sleeping or remaining still. By raising their body temperatures, birds can survive minus fifty-degree Fahrenheit readings.

ADDITIONAL READING

Bellrose, Frank C. *Ducks, Geese, and Swans of North America.* Mechanicsburg, Pa.: Stackpole, 1976.

Bolsenga, Stanley J., and Charles E. Herdendorf. *Lake Erie and Lake St. Clair Handbook*. Detroit: Wayne State Univ. Press, 1993.

Costello, David F. *The World of the Gull.* Philadelphia: Lippincott, 1971.

Pasquier, Roger F. *Watching Birds: An Introduction to Ornithology.* Boston: Houghton Mifflin, 1977.

Landbird Fallout

In the spring of the year the small birds of passage appear very suddenly . . . which is not a little surprising, and no less pleasing: at once the woods, the groves, and meads, are filled with their melody, as if they dropped down from the skies. The reason or probable cause is their setting off with high and fair winds from the southward; for a strong south and south-west wind . . . never fails bringing millions of these welcome visitors.

—William Bartram

Northern parula warblers are active foragers that hover at the ends of branches and may hang upside down like chickadees. As nesters they are more common in southern states and in Maine and the Maritime provinces. In the south they nest in Spanish moss, in the north in hanging lichens. Few linger around Lake Erie except sparsely on the southern shore, though they may be seen as uncommon migrants to northern Ontario. The yellow breast and dark breast band of this spring male are obscured by foliage, but his white "eye shadow" is clearly visible.

(*Opposite*) Black-billed cuckoos winter in northern South America south to Ecuador, Peru, and Bolivia and summer in North America mostly east of the Mississippi. They are less numerous than their close kin the yellow-billed cuckoo. Cuckoos are especially fond of hairy caterpillars such as gypsy moth larvae and are more common at the site of outbreaks. Curiously, they can shed their stomach linings to rid themselves of the bristly hairs. Though European cuckoos regularly lay their eggs in other birds' nests, American species of cuckoo rarely parasitize, and then only each other. Their nests are flimsy affairs, at best.

Many thousands of modern birders share the same thrill William Bartram felt two centuries ago as he watched the great surge of songbirds that invades this region in April, May, and early June. Whether one is an expert birder on the watch for rarities or a novice captivated by the first sighting of a luminous scarlet tanager, a "wave day" in May is not likely to be forgotten.

Anyone can share in the magical rush of spring migration by going outside and opening both eyes, even in city green spaces. However, experienced birdwatchers know that Lake Erie weather and geography create "hot spots" where songbirds—thrushes, tanagers, warblers, flycatchers, grosbeaks, and others—mass in exceptional numbers and variety on certain enchanted days.

Migration itself is an immensely complex phenomenon, and explaining in depth the how's and why's of it is well beyond the scope of this chapter. There is no simple map or starting gun for the clouds of birds that fly south in late summer and autumn and

Here a male summer tanager eats a hornet. Tanagers are bee eaters, and this one is sometimes called "red beebird" in the south where it commonly nests. Summer tanagers do not usually nest north of southern Ohio and Pennsylvania. This one, photographed on the Pelee side of Lake Erie, is an "overflight" and is one of the spring migration's moments of serendipity.

Warbling vireos are common, especially in upstate New York and southern Ontario, but are more often heard than seen. Migration fallouts caused by changeable weather conditions are a boon to birdwatchers; they may force birds like this one out of the treetops, where they usually lurk, to lower levels for good viewing. Though warbling vireos are sometimes mistaken for warblers, the beak is heavier, and the broad white eye stripe is another field mark.

press north again from warmer climes to breed. In this hemisphere, some migratory birds winter in more clement parts of North America. Many songbirds, however, fly south to the West Indies, Central America, and northern South America and are called "neotropical migrants."

Ornithologists argue over how migration developed, but they are in general agreement about its advantages. Though the long trip from the wintering grounds to the summer range and back is fraught with danger for individual birds, it bestows a reproductive edge on species. North America offers burgeoning insect populations in spring and summer, and many birds, especially shorebirds, may fly clear up to the Arctic tundra to breed. This is the concentrated protein young birds need to grow and develop. There is also less competition from other birds up

Rose-breasted grosbeaks are common residents in the Lake Erie area. The flashy costume of this male contrasts with the drab brown of his mate, who looks rather like a large sparrow. These grosbeaks winter from the West Indies and Mexico to northwestern South America and, along with many other songbirds, are termed "neotropical migrants." Much concern has recently been expressed about these birds' fates in relation to the clearing of subtropical and tropical forests.

A true jewel of spring and summer forests, this male scarlet tanager is another common migrant in the Lake Erie region. Females, immatures, and adult males in winter are dull green above with dark wings and yellow underparts. While the rose-breasted grosbeak's song has been likened to "a robin who has taken singing lessons," the poor scarlet tanager sounds like one with a sore throat.

Black-throated green warblers are some of the more common migratory warblers in this area. Note this male's yellow face, black throat, and dull green crown. Females have a less impressive black cravat. Nesters are more common in Pennsylvania, the Adirondacks in northern New York, and Ontario east of Georgian Bay than right around Lake Erie. This is because they prefer coniferous or mixed coniferous and deciduous forests in their summer range.

north than there is in the tropics and semi-tropics, and the midnight sun provides longer foraging hours. Migrating may be dangerous, but in the game of avian evolution, the payoff is worth the risks.

A small bird is not a windup toy that leaves its wintering ground and motors on to its nesting range on automatic pilot. True, many aspects of bird migration—such as general direction and distance to be flown—are programmed genetically. For example, many first-year birds make the southward journey from their birthplaces in the north without prior experience or even the company of their parents. However, if a fledgling is to live long enough to reproduce, its behavior must be flexible enough to cope with geographical barriers and shifts in wind, temperature, and other aspects of our changeable and often dangerous weather. To survive, it must learn to make decisions, and on each of these its life may truly depend. Migration is strenuous and exacting work.

What combinations of environment, instinct, and strategy bring these showers of colorful passerines to Lake Erie, and when and where can birders find them? First, birds must be able to locate food, and quickly. Some of the earlier migrants, such as yellow-rumped and palm warblers, can subsist on last year's remaining fruits and berries until the first big hatch of insects emerges. Others, like thrushes, may probe for worms until insects appear. Most May sojourners, however, have followed the hatch north, one that coincides with opening leaf buds on which the insects themselves feed.

Scientists think that a bird becomes restless and ready to migrate through the interplay of environmental stimuli such as changes in day length and in some cases the workings of an inborn yearly clock (changes in day length do not occur at the equator). Shifts in metabolism at this time enable birds to store masses of fat very rapidly under their skins and around their vital organs. Fat is the fuel for flight, though a bird that has used up its reserves may start burning muscle protein as a last resort. A small bird can travel about 125 miles on one gram of fat.

Once birds have begun their migratory flights, their strategies are as numerous as the species themselves. Some fly by day; some, including most songbirds, by night. Some readily cross large bodies of water, and some, such as blue jays and hawks in the buteo group, will not. Certain birds, like swallows, which swoop after flying insects, feed and migrate simultaneously. Others, like the well-known American golden plover and blackpoll warbler, fly thousands of miles over open ocean on

Cerulean warblers usually nest south of Lake Erie, though some travel to Ontario east of the Georgian Bay. Although they appear to be extending their range north and east from centers in West Virginia, eastern Kentucky, and southern Ohio, they are in trouble because they will nest only in deep riverside woods and are very sensitive to fragmentation of breeding habitat. Watchers usually have to crane their necks to see these beautiful creatures, but during migration the birds may literally be at their feet.

A handsome male Cape May warbler shows off his rich chestnut cheeks contrasted in yellow. As with most warblers, the male is much showier than the female, who, in this case, is duller and lacks the chestnut coloring. Since this species nests in spruces, members fly well north of Lake Erie to raise their young and are attracted by outbreaks of spruce budworm. Ironically, one can see a Cape May at Cape May, New Jersey, only during migration.

Birdwatchers try for a big day on the elevated boardwalk at Magee Marsh west of Port Clinton on the south shore of Erie's western basin. This and Point Pelee National Park on the north shore are two of the best birding locations in North America. Here on the boardwalk and in the neighboring marsh, birders can hope to see a hundred species of birds in a day, including perhaps twenty-five warbler species.

their autumn flights, relying on their fat reserves and tail winds to carry them all the way from Canada's maritime provinces to South America. As a whole, species follow established patterns of migration, and birds are surprisingly faithful to both their migration rest stops and summer nesting places.

Edwin Way Teale estimated in *North with the Spring* that the season advances about fifteen miles a day from its beginnings in the far south to its blossoming in the so-called temperate regions. However, most birds' progress is not such a steady one. William Bartram was right about the importance of "fair winds from the southward." Birds rest, eat, and stage until they can catch a promising warm front from the south, and then they move in waves. There are several reasons for this. First, such a front promises birds a healthy hatch of insects, the fast food they will need to restore shrinking fat reserves. Second, a tail wind increases birds' air-to-ground speed and allows them to cover more distance on less fuel. Clear nights also help. Passerines usually fly at night for several reasons: night flight frees the daylight hours for foraging; because the cool air helps dissipate body heat, there is usually less air turbulence then, and they can navigate by the stars (among other methods). They are also safer from predators at night.

Spring migration lasts for at least four months, activity at the western end of Lake Erie usually two to three weeks ahead of that at the other end to the northeast. Conversely, it is two weeks later in autumn. In early March the first wave of raptors, especially harriers, eagles, and the larger buteos like red-shouldered hawks, arrives at the lake. Horned larks and crows—the real early birds—have been moving for some time already. Aldo Leopold's "goose music" resounds through the marshes, and flights of tundra swans pull in from the Chesapeake.

Other waterfowl begin to come as well. And pugnacious bachelor red-winged blackbirds are setting up territories in the snow-covered marshes hoping to attract females soon to arrive, and the tinkly voices of song sparrows grace the yet-frigid dawns. Around the Ides of March, turkey vultures make their celebrated return to Hinckley, Ohio, rehearsing the role of Shakespeare's bird of ill omen in *Julius Caesar*. They are accompanied to Lake Erie by buteo hawks, accipiters, and shorebirds. Later in the month all kinds of waterfowl will be on the move, and many will rest and feed in the marshes and Lake Erie shallows through early April.

Though it nests only rarely around Lake Erie, the white-throated sparrow is abundant in the mixed hardwood and especially the boreal forests farther north. In New England it sings "Old Sam Peabody, Peabody, Peabody," but north of the border it says, "My Sweet Canada, Canada, Canada." During migration, birders on Lake Erie can see hundreds on a good day.

This wood thrush is digging for worms as it waits for more insects to hatch. It is not as affected by cold or a late insect hatch as many migrants. Wood thrushes are fond of fruit and have been reported to feed at least some to their nestlings. They like deep, moist deciduous woods, and their lovely songs have been decreasing in North America as a result of the forests' fragmentation.

Baltimore orioles, like thrushes, may resort to digging for worms until insect populations grow. The year 1997 was a good year for orioles in this region: Spring was so late that many stayed to nest near the lake rather than continuing farther north. Orioles' genus name comes from the Greek word *ikteros*, meaning jaundice. Did the word refer to the birds' color, or was it thought that seeing one was a cure for the condition?

Like Cape May warblers, black-throated green warblers, and white-throated sparrows, many veeries move north of Lake Erie to nest in Canada. This thrush likes wet, second-growth deciduous woods where its beautiful downwheeling song—described by some as metallic, by others as liquid and breezy—is commonly heard. It usually searches for insects on the ground and nests on or near it.

(*Opposite below*) Wintering from the southern Atlantic and Gulf Coasts south to Honduras, Nicaragua, and Costa Rica, tree swallows migrate around the Gulf of Mexico (as do other swallows) rather than across it. These pretty birds are tougher than they look, and they arrive in our region earlier than other swallows do. Here a flock of them hunkers down to wait out a cold front. Common nesters from Ohio and Pennsylvania into northern Ontario and Quebec, they use tree cavities or nest boxes, preferably near water.

Songbirds are moving in April, too, especially the ones headed for far-northern Canada. The first of the great waves brings kinglets, juncos, tree and song sparrows, winter wrens, and brown creepers to the edge of the lake. At the month's end and in early May, a surge of yellow-rumped warblers, ruby-crowned kinglets, white-throated and white-crowned sparrows, and hermit thrushes appears, as well as black-and-white and palm warblers, flickers, and blue jays. This is a good time to look for a blue-headed vireo. At the end of April, sharp-shinned hawks come through, although not in the numbers they will show in the fall migration. The bulk of the birds arrive in May: whippoorwills, hummingbirds, warbling vireos, magnolia warblers, American redstarts, black-throated blue warblers, orioles, and least flycatchers. Later come yellow-bellied flycatchers, blackpoll and bay breasted warblers, more flycatchers, and waxwings. By the end of May, the great show is usually over, except in very late years.

These birds and the peak of the migration in May can be viewed from many sites on both the southern and northern shores of the inland sea. In a shotgun effect, bunches of birds may visit one place but be absent from another. In addition, Lake

The northern mockingbird is an uncommon migrant and resident in the region. It is slowly expanding its range northward. A wide range of open and partly open habitats suits its needs, and it is often abundant in suburbs farther south. Mockingbirds feed on berries, invertebrates such as sow bugs, and the occasional small vertebrate. Their adaptability suggests that they will increase in this area if winters remain mild.

A common migrant, this black-throated blue warbler takes a moment's rest from insect foraging at Point Pelee before resuming its flight north with the spring. However, the curving shoreline of Lake Erie appears to disorient some birds, and they may find themselves heading south again to Pelee's tip. This phenomenon is called reverse migration.

Erie's geography concentrates migrants at certain spots legendary among international birders. The most celebrated of these are in the western basin—Point Pelee on the north shore in Ontario and Magee Marsh on the opposite coast in western Ohio.

Why these sites particularly? For one thing, they are strategically located at the crossroads of two major North American migration routes, branches of the Mississippi and Atlantic flyways. For another, they are placed so that they funnel together birds that are trying to deal with a formidable water barrier. They also provide mature forest, marsh, and thickets for insect foraging in the midst of what is increasingly an agricultural desert, from a bird's point of view.

For waterbirds such as gulls, terns, and ducks, the lake can be both a highway and a staging ground for rest and feeding. For landbirds, however, it can literally become a watery grave. Songbirds like warblers are quite reluctant to cross the gray-blue water because if the weather changes and the wind turns, if it rains, or if turbulence blows them off course, they may never reach the Canadian shore. This means that if William Bartram's "high and fair winds from the southward" fail to materialize, a kaleidoscope of small birds lingers on the south shore of the lake until conditions improve. If weather has been cold, with northerly winds blowing for some time, concentrations on the bird trail at Magee Marsh west of Port Clinton, Ohio, may burgeon once the winds shift.

Birders should watch nightly weather reports for low pressure cell locations, anticipated temperatures, and wind directions. In early spring, southwest winds and clear skies are important for raptors. Later, when the songbird waves begin, a low pressure cell in the Arkansas region and expected temperature jumps are good predictors. Warm nights are better than cool ones. In some years, the weather is more cooperative (for humans) than in others. When it is not, migration can be very subtle, because a series of slow-moving fronts may allow birds to fly above or around them, disappointing birders waiting for a major wave. An early spring in which vegetation leafs out earlier farther south enables many birds to gather food as they migrate, and they will not have to stage by the lake in large numbers. Leaves also make them harder to see.

If the weather is not ideal for travel, birds rest, feed, and wait for better conditions at the Magee Marsh bird trail. Although small, the area has good foraging habitat—more than 150 species of songbirds and thirty-eight warbler species have been recorded here where a raised boardwalk makes the area accessible to virtually anyone who loves birds. Magee Marsh and nearby Crane Creek State Park were haunts of my parents, both enthusiastic birdwatchers, and I cherish many memories of tree swallows, great-horned owl nests, basking turtles, and warblers in vivid spring plumage from our repeated visits there.

Many birds filter around the western or eastern ends of the lake after they have fed and rested. Most of them circle west, but dark-eyed juncos and some warblers may round Erie's eastern

Yellow-rumped warblers are early migrants that arrive here in good numbers toward the end of April. Although some winter in the southern United States, many others fly farther south to Mexico and Central America. During the past twenty years, ornithologists have warned about deforestation and its effects on our neotropical migrants. This is a serious concern, although the picture is somewhat more complex: Surveys by the Ohio Department of Natural Resources in the mid-eighties showed that although about one fourth of neotropical migrants had declined, one third remained stable, and 40 percent actually increased.

end. When the wind shifts to the southwest, however, other tiny adventurers seize their opportunity and take off from the south shore across the lake itself toward Canada. At Point Pelee near Leamington, Ontario, even greater concentrations and varieties of birds can be seen than those staging on the south shore. This remarkable area has recorded about 365 bird species, of which warblers account for more than forty. It is often considered one of the top ten birding locations in North America. Avid birders gather at the tip of the Point Pelee peninsula at dawn to see which weary avian travelers will appear out of the morning haze. These birds have often followed the chain of islands that stretches from north of Port Clinton and Sandusky, Ohio, toward Canada: South, Middle, and North Bass islands, Pelee Island, and the smaller fragments of the archipelago.

The largest number of migratory birdwatchers wait at Pelee between about May 10 to May 15. Park records show that approximately 80,000 birders flock to the peninsula in that month each year, a major boon to the area's economy. (Large banners pronounce "Welcome Birders" on the road to the waterfront in Leamington.) For those who dislike crowds (of

This savannah sparrow was photographed at Pennsylvania's Presque Isle State Park. One of the finest birding sites in Erie's eastern basin, Presque Isle's miles of sand beach, sand plains, and wetlands attract many spring and fall migrants. Savannah sparrows are common in grasslands, marshes, and other open habitats.

humans, that is), Pelee may seem something of a mob scene. Mega camera rigs abound, as well as state-of-the-art spotting scopes. Television camera crews ask men with big rigs and birding hats questions like "What dangers lurk out there for you birders?" They seem a bit disappointed to hear the un-macho replies: "Mostly sunburn and poison ivy. Actually, the main dangers are to the birds, through habitat destruction." Few people are seen consulting bird guides, the assumption being that one has already memorized them thoroughly.

Yet there is a lively sense of comradery here. Experienced birders help novices with identifications. Knots of people gather to see rarities spotted by experienced—or lucky—watchers. Their eyes shine. I remember with special affection the couple who pointed out a golden-winged warbler to me and my husband.

"Seeing it was an accident. We were just walking back to our truck. It's a lifer for us—isn't it beautiful?" they beamed.

This is the time for gaudy tanagers, grosbeaks, and orioles, birds that seem to be everywhere and that send an ordinary birdwatcher's heart to fluttering. The experts respond with gentle tolerance when one points them out; bird mavens are after rarer game. Females of ruby-crowned kinglets, white-throated sparrows, yellow-rumped warblers, and Swainson's thrushes also abound, as well as more unusual species such as blackpoll and mourning warblers. Two sandhill cranes far from their usual territory fly over, necks outstretched, calling with voices that sound like the prairies. Around Memorial Day another wave of birds sweeps in, featuring American redstarts, indigo buntings, vireos and flycatchers, and many female warblers. By mid-June the crowds of both birds and birdwatchers have thinned, with the birds nesting and raising the next generation of migrants, and the birders turning their attention to the breeding birds and beginning to anticipate shorebird migration.

It's important to note that Point Pelee and Magee Marsh are not the only good places to witness the spring migration drama.

A "chebec" call tips off birders that this is a least flycatcher, one of the five empidonax species to be expected during migration. These small flycatchers are very hard to identify visually, but their respective calls give them away. Least flycatchers, the earliest migrating empids, often arrive at Point Pelee in late April.

In fact, for those who prefer to avoid human crowds, there are Hillman Marsh and Wheatley Provincial Park near Pelee, Rondeau Provincial Park about fifty miles east of Pelee on the lake's north coast, and Long Point, yet farther east on the same shore near Port Rowan. Sheldon Marsh and Mentor Headlands in Ohio and Presque Isle near Erie, Pennsylvania, on Erie's south shore, as well as other sites listed in this volume's appendix, also afford exciting and most satisfying spring birding experiences. Of course, any of these areas will be less populated during the week than on the weekends, and it is certainly wise to visit Pelee during midweek, if possible.

What many ardent birders hope to see, at least someday, is not just an ordinary wave day but what is called a "fallout day." Fallouts happen when the weather is cold, especially at night, and birds are short of food. If a warm front carrying a wave of birds meets a cold front from the north over Lake Erie, the warmer air slips over the cooler, forcing them to fly higher with greater effort. Then it cools down itself. The wearied birds must eventually descend, and the sight of the green spike of Point Pelee thrusting into the perilous waters attracts tenfold numbers of them. If rain accompanies the fronts' impact, the birds are grounded and, as Tom Powers notes in *Great Birding in the Great Lakes,* "The birdwatching takes a quantum leap beyond incredible" (146–47). Birdwatchers thrill at the opportunity to see unusual species in double rather than single digits. One birder was heard to say, "It's just like looking at a bird book, only everything is alive!"

Major fallouts usually take place early in May when the weather is less settled than it is later in the month. The later in May a fallout occurs, the greater variety of species can be seen; a major fallout may happen at this time every four or five years. Small birds can be so exhausted by the time they reach Pelee that they sometimes even land on birders standing at the end of the point. Some, scarcely more than skin, feather, and bone, may be too tired to feed and will die if they cannot. Sometimes they are even seen picking dead insects from spider webs and from the radiators of cars, as was the case during a big warbler die-off in 1996. If the trees have not yet leafed out, the gamble to be first to the nesting grounds can turn deadly. However, if birds make it from the beach sand to the vegetation beyond, and if conditions are reasonable, they will find a diverse habitat of forests, wetlands, and grassy areas in which to forage and recoup their strength for the next leg (so to speak) of their long flight.

It is an irony that birds' struggles during a fallout can result in so much human pleasure, as Ada Clapham Govan noted in her 1940 book *Wings at My Window:* "Easter Sunday was a mellow, sunny day, but a blizzard broke that night, catching the migrating hosts in their northward flight and slaying them by thousands and tens of thousands. The noise of that wild storm and a mixed chorus of bird calls outside awakened me at daybreak. Every feeding box was inches deep in snow; yet birds were everywhere, and in spite of the storm, they were singing lustily, joyously, as I had never heard birds sing before" (Eriksson and Pistoriono, 14).

Migrating benefits species that journey from the tropics to raise their young amid the burgeoning summer insects and lesser competition from other birds in the north. But it is fraught with danger for individuals, especially when storms hit without warning. Conflicted as we may feel about the great mortality such gales cause, we who watch these small travelers are the inheritors of joy from their short and strenuous lives. Certainly, avian migration is a major wonder of our natural world, blessing human beings from William Bartram to those of us who watch the "millions of these welcome visitors" around Lake Erie today.

ADDITIONAL READING

Ehrlich, Paul R., David S. Dobkin, and Darryl Wheye, *The Birder's Handbook.* New York: Simon and Schuster, 1988.

Eriksson, Paul S. and Alan Pistorino. *Treasury of North American Birdlore.* Middlebury, Vt.: Paul S. Eriksson, 1987.

Goodwin, Clive E. *A Bird-finding Guide to Ontario.* Toronto: Univ. of Toronto Press, 1995.

Kerlinger, Paul. *How Birds Migrate.* Mechanicsburg, Pa.: Stackpole Books, 1995.

Peterson, Roger Tory. *Eastern Birds.* New York: Houghton-Mifflin, 1980.

Powers, Tom. *Great Birding in the Great Lakes.* Flint, Mi.: Walloon Press, 1998.

Area Nesters

The witnessing of this "territory establishment" sealed my fate for the next seven years. I was so fascinated by this glimpse behind the scenes with my Song Sparrows, that I then and there determined to watch Uno for several hours every day, so as to follow the daily course of his life, to find out the meaning of his notes and postures, in short, to discover exactly what he did and how he did it. In particular I wanted to study the matter of "territory."

—Margaret Morse Nice, *The Watcher at the Nest*

As the first days of June wane, flocks of birds and birdwatchers thin at Point Pelee, Crane Creek, and other Lake Erie migration hotspots. Many birds have passed on to northern forests and Canadian tundra; many birders have gone home to cultivate their gardens and dream of fall migration. Perhaps they'll brush up on how to know the shorebirds already about to trickle back from the Arctic at the end of the month, to distinguish raptors in flight, or to recognize warblers' confusing fall plumages.

Birds that nest in the north country are now breeding there, fueled by the extraordinary fertility of black flies, mosquitoes, and other insects on the nesting grounds. It's a comfort to realize how many blood-sucking insects end life each year in the crops of the birds' hungry young! Other avian species, however, remain in our own area's woodlands, fields, and marshes, busily absorbed in reproducing their kind. This is, after all, the reason many of them braved distance and weather to come here.

Seen from a visiting spaceship in 1800, Lake Erie must have looked rather like an island of water in a nearly uniform green sea of broad-leaved trees: Of the five Great Lakes, it is the only one to lie completely surrounded by the Eastern Deciduous Forest Zone. This fact goes a long way to explain why some birds migrate well north of Lake Erie to nest while others stay to enliven our early summer mornings and evenings with their busy

(*Opposite*) The willow flycatcher likes old fields and other open areas. It gives its call in early July in sparse thickets such as sumac. This flycatcher is much more common than its virtually indistinguishable cousin, the alder flycatcher. The two species were once lumped into one, Traill's flycatcher. As is true with other small empidonax flycatchers—least, Acadian, and yellow-bellied—birders can best distinguish these two by their songs.

goings-on. It also explains why, despite local differences and human influences, the range of habitats around Erie is fairly consistent.

Not far to the east, in the state of New York, begins the hardwood transition forest (also called Alleghanian) that covers the Appalachians except at their highest elevations. Not far north of Lake Erie in Ontario begins a similar transition forest. This zone of mixed hardwoods and evergreens grades from the deciduous woodlands immediately north of the lake into the vast boreal forest, that dark-green zone blanketing Canada north of the Great Lakes and south of the Hudson Bay lowlands and the Arctic tundra.

Birds adapted to nesting and feeding in spruce, fir, and pine naturally pass through the Lake Erie region rather than lingering to breed here (nesting preferences of American wood warblers, for example, are particularly specialized). Many of these birds fly to the hardwood transition forest, where white and red pine and

Graceful barn swallows have adapted well to life near humans. They are common around bridges, barns, and other outbuildings, where they glue their mud nests in sheltered spots such as under the eaves. They often appear hawking for insects above the surface of farm ponds and park lakes. Their eggs, white spotted with brown, lie on thick beds of feathers, and the birds usually produce two broods a season.

The red-eyed vireo is a bird of the midcanopy in deciduous forests, its nest a well-made cup suspended from a forked twig. Materials include vines, rootlets, fine grass, and the amazingly strong spiderwebs used by many other birds as well. Red-eyed vireos are among species most heavily parasitized by cowbirds, which lay eggs in vireos' nests and crowd out the rightful young. Once extremely abundant, red-eyed vireos are still quite common but are declining along with cerulean warblers, which have been hurt by the loss of bottomland forests. Warbling vireos and brilliant scarlet tanagers also reside in the deciduous canopies.

hemlock mix with yellow birch, sugar maple, American beech, and other deciduous trees, or they travel farther to the spruce- and fir-dominated boreal zone. Birds like white-throated sparrows, black-throated blue warblers, blue-headed vireos, hermit thrushes, and dark-eyed juncos breed in the Ontario hardwood transition zone and farther north, though some juncos may nest in cool hemlock ravines south of Lake Erie in Ohio, Pennsylvania, and western New York. A few northern warblers do so, too. These and other northerly breeders often summer southward in the mixed hardwood transition forests along the cool flanks of the Appalachians as well. Other types of birds—shorebirds especially—leapfrog all the way to the Arctic tundra of the far north.

American redstarts, like vireos, are frequent cowbird hosts, and they will occasionally build another floor above the cowbird eggs in the bottoms of their nests. They are fond of river valleys, mostly on Lake Erie's southern shore, nesting low in trees and sometimes on the ground. Rose-breasted grosbeaks and indigo buntings also frequent riparian corridors. When redstarts forage for insects, they vigorously flash their tail and wing patches, hence their name.

The nesting birds in our deciduous zone represent both those adapted to the thick, broad-leaved forests of the pioneer past and those suited to altered conditions of human settlement and development. Anyone who studies the history of bird life in Ohio, Michigan, Ontario, New York, or Pennsylvania will be struck by the speed with which bird populations have shifted in response to human influences, have waxed and waned and (sometimes) waxed again over the past two centuries.

BIRDS OF FOREST INTERIORS

Ironically, species from two opposing groups—birds adapted to the primeval forests and those that invaded the area because settlers cut the forests—are those most likely to be declining around Lake Erie now. Today it's hard to imagine the speed with which forests melted away in the nineteenth century. Before 1850 woodsmen had cleared 12.4 million acres in New York, 8.6 in Pennsylvania, 9.8 in Ohio, and 1.9 in Michigan. In the ten years before the U.S. Civil War, the process accelerated, and in that one decade 2.1 million more acres of forestland disappeared in New York, 1.9 in Pennsylvania, 2.8 in Ohio, and 1.5 in Michigan. Farmers would live to regret the big piles of logs they burned before wood became scarce and expensive.

Unfortunately, the limited strip of mixed deciduous forest on the north shore of Lake Erie in Ontario (termed "Carolinian forest" in Canada) coincided with Canada's richest agricultural land and mildest climate. Most of it is gone now. The small patches remaining of this diverse forest stand on flood plains and swamps that resisted farming: Point Pelee, Rondeau Provincial Park, Long Point, Backus Woods, Springwater Woods, and Catfish Creek. Other remnants line stream gullies that empty south into the lake. All these provide important nesting habitats for birds.

Lark sparrows are western birds whose populations expanded around Lake Erie at the turn of the last century. Although they once ventured as far east as West Virginia and western Pennsylvania, Maryland, and Virginia, they are now a relict species in the Lake Erie region. The last confirmed breeding in southern Ontario was in 1976, and they apparently do not breed in either New York or Michigan now. Perhaps fifteen pairs nest in Ohio, where they are found especially in the old sand areas of Kitty Todd Preserve and Oak Openings Metropark west of Toledo. The preserve must be specially managed to provide the sparse ground cover lark sparrows prefer.

Naturally, the forest clearances affected deep-woods birds profoundly. This was especially true of species limited to woodland interiors and ill adapted to forest edges: broad-winged hawks, barred owls, Acadian flycatchers, black-throated green, cerulean, hooded, Canada, and black-and-white warblers, Louisiana waterthrushes, ovenbirds, wood thrushes, and others.

OPEN-COUNTRY BIRDS

It's an ill wind that blows nobody good, however, and the forest clearings increased nesting habitat for grassland birds. Many of these invaded the land around Erie from the west as agriculture spread. For example, the horned lark, a bird of the western plains, was first reported as an Ohio nester about 1880. Bobolinks were first seen around the west end of Lake Erie in 1872, and savannah sparrows showed up about the turn of the twentieth century. Eastern meadowlarks, vesper sparrows, upland sandpipers, and dickcissels also appeared in the farmlands. Sedge wrens and Henslow's sparrows were attracted to dense, low grass. (They nest in this region only occasionally now.) Grasshopper sparrows appeared locally in the sparse grass of quarries. Barn owls cruised the new hay fields and meadows in search of their staple food, meadow voles; kestrels and red-tailed hawks replaced woodland raptors such as sharp-shinned, Cooper's, red-shouldered, and broad-winged hawks. American crows supplanted ravens. More recently, turkey vulture populations have exploded in Ontario during the past thirty years.

Some of these birds, those that did not need wide expanses of grassland, may have lived in forest openings before settlement, but their numbers skyrocketed when the land opened up. From about 1815 to 1910, farming created much fence-row and thicket habitat, which supported dense bird life. Louis Campbell, in *Birds of the Toledo Area,* estimates that numbers of birds were probably at their peak in that area about 1900.

This trend reversed in the next century, and open-country birds have declined in their turn: Hay crops like timothy and clover dwindled as cars replaced horses, and row crops like corn and soybeans supplanted grasslands and meadows. When the "clean farming" movement began about 1935, fields grew larger

(*Opposite*) Along with other grassland nesters, eastern meadowlarks spread into the area as the countryside shifted from forests to farmland. Unlike many open-country birds, which have declined sharply in recent years, meadowlarks are still fairly common in areas around Lake Erie. Males can be seen and heard defending their territories from tall weeds or other vantage points with clear "spring-of-the-year" whistles.

Among the first breeding birds to arrive in spring and among the last to leave in autumn are killdeers, very successful plovers. Like field, song, and chipping sparrows and common yellowthroats and goldfinches, killdeers breed in many habitats; they have even been observed nesting between railroad ties. Their buff-colored eggs, laid in a shallow soil scrape, are black speckled, making them extremely hard to see among small stones and pebbles. Adults are very visible and vocal (their scientific name is *Charadrius vociferus*); however, they are canny when they leave their nests and use broken-wing displays to lure predators away from their two broods a year.

to accommodate mechanical cultivation, and farmers cropped them right up to their edges, destroying the rich fence-row and thicket habitat of earlier times. This was especially true on the flat, fertile soils of northwestern Ohio and southwestern Ontario.

Elsewhere, farms on marginal land failed in increasing numbers during the first half of the twentieth century. Inevitable plant succession began on the abandoned land, with brushy old fields and second- or third-growth woodlands replacing habitat that had fostered grassland birds. Bobolinks are now uncommon in the area, and dickcissels have retreated to the western end of the lake, though they make sporadic irruptions into southwestern Ontario and Ohio. Eastern meadowlarks, though still relatively common, are in decline, and Henslow's sparrows have decreased sharply in the past thirty years. Perhaps five pairs now breed in Ontario. Barn owls are extremely rare in the region, both in the states around the lake and (as they have always been) in southern Ontario. The loggerhead shrike, another predator of open

Though many people think of sparrows simply as LBJs—little brown jobs—their plumage is quite attractive in good light. The song sparrow is one of our most common area breeders, beginning to sing while snow is still on the ground and using a wide variety of habitats, including urban areas. Its adaptability and high birth rate of two or three broods, and even occasionally four, help account for its success.

country, is listed as threatened both north and south of the lake and has not bred in the immediate area for many years.

BIRDS OF OLD FIELDS AND SECOND-GROWTH WOODS

As plant succession on old, abandoned fields progressed, another group of birds flourished. These included eastern kingbirds, brown thrashers, gray catbirds, common yellowthroats, song and field sparrows, yellow warblers, indigo buntings, yellow-breasted chats, and, sporadically, northern bobwhite quail. Many of these species are still common in the area today, especially those that are adaptable nesters. But nature never stands still: Some farmers left the land as much as a century ago, and the old fields are aging. This means that many of them are now woodlands again, although woodlands of different ages and compositions than the original "forest primeval." Some of the old-field birds began thinning. Breeding bird surveys show declines or range shrinkage in species like eastern kingbirds, brown thrashers, and yellow-breasted chats, though these birds are sometimes fairly common locally.

Certain species, however, such as American redstarts, Nashville warblers, blue- and golden-winged warblers, and chestnut-sided warblers prefer early-stage woodlots rather than old fields, and some are increasing here today. Chestnut-sided warblers, which are fairly common in brushy edges and young woods in Pennsylvania, New York, and Ontario, have recently expanded farther into Ohio, at the southern edge of their breeding range. So have veeries, thrushes that are fond of young, moist woodlands near streams.

While maturing timberlands might seem to predict a resurgence of our original deep-woods birds, the picture is not so simple. Pictures seldom are. In some parts of the region, forest percentages are certainly a good deal higher than they were a hundred years ago. However, these areas are mostly on the edge of Appalachia or north of the Carolinian forest zone in Canada and not closely adjacent to the lake. In heavily farmed southwestern Ontario, southeastern Michigan, and western Ohio especially, woodlots continue to disappear. In Ontario's Essex and Kent Counties near Windsor, for example, only 5 percent of the original forest cover remains.

Even in areas where trees have regrown, woods are much more fragmented than they were two hundred years ago and are often islands in the midst of a sea of agriculture, suburban development, and highway networks. A large hawk or owl may need at least a square mile to establish a hunting territory, an area that is now hard to find in many places; also, small woodlots are much less diverse than large ones, in general. Species may have a hard time dispersing to other woods as well, because of problems crossing unsuitable territory between the forest fragments.

(*Opposite*) Nesting from ground level to about three feet above ground, the common yellowthroat has double broods, unlike most other warblers. Although yellowthroats are prime victims of cowbirds, they have remained abundant. Wetland birds, they also nest in fields, where their well-camouflaged nests and caution in coming and going help keep their populations high.

Also working against certain kinds of birds is the so-called "edge effect." Species loss is much higher in small patches of woods than it is in ample timberlands. Besides denying birds adequate territories, fragmentation leaves their nests more vulnerable to raccoons, grackles, jays, and other predators that hunt the woodland margins. Nests near the "edge" also draw brown-headed cowbirds that parasitize many birds, particularly flycatchers, thrushes, tanagers, warblers, and vireos. Originally, cowbirds probably lived in open country west of the Mississippi, land that was suitable for their ground-feeding habits and courtship displays. They ventured east as settlers cut the virgin forests, but they were not very common until this century, when their populations ballooned. Females lay eggs in other birds' nests, and the aggressive, fast-growing cowbird young crowd out their foster siblings. Many eastern species of birds have encountered cowbirds only recently, and they have not had time to evolve defenses such as tossing out the spurious eggs or building new nests, as "experienced" species have learned to do. The wood thrush, extremely sensitive to forest fragmentation and cowbird parasitism, has steadily declined over the past twenty years.

Still, the area has seen increases in some forest birds, such as pileated woodpeckers, which are fairly common in mature woodlands and ravines south of Lake Erie; they are local north of the lake in Canada. Yellow-bellied sapsuckers and red-headed woodpeckers, birds that like shaded openings in moist woods, are local both north and south of the lake, though redheadeds have declined significantly in this century. Northern flickers are common nesters throughout our woodlands.

SOUTHERN INVADERS

Some birds' ranges have pushed north during the past century. Best known in this group are northern cardinals, year-round residents whose bright feathers and buoyant songs enliven summer and winter alike. These pretty birds were nowhere to be seen in the Lake Erie region two hundred years ago. They apparently spread to the lakeside counties of Ohio in the 1830s, but breeding pairs were still rare at the end of that century. On the Niagara Frontier in New York, they appeared only in the

1880s. Numbers rose strongly in the first decades of the twentieth century. Today's bird guides show cardinals ranging to the northern boundary of the Great Lakes and to New Brunswick and Nova Scotia to the east.

Other southerners in our area include the tufted titmouse, now common on the south shore of the lake but rare in Ontario and largely confined to the Carolinian forest zone there. In Canada it is most often found on the Niagara Peninsula and was

Eastern bluebirds nest in old woodpecker holes and other natural cavities when they are available. Thus, they were hard hit when wooden fence posts that supplanted old-growth trees were themselves replaced by metal posts. Booming populations of hole-nesting house sparrows and starlings also compete with the less aggressive bluebirds for nest sites. In recent decades, however, bluebird nesting box programs have helped these beautiful birds to recover.

first reported there in 1914. Kentucky warblers have modestly increased their range northward in recent decades but are not regular breeders in Ontario. Orchard orioles are moving into our area from the south and west, and prothonotary and hooded warblers and Acadian flycatchers have also expanded their ranges north, but are very rare on the Canadian side of the lake. The same is true of northern mockingbirds.

Carolina wrens, fond of thickets and tangles, also reach the northern edge of their breeding grounds in southwestern Ontario. The three extreme winters from 1976 through 1978, which affected so many species, extirpated these wrens from the Erie region and thinned their ranks throughout eastern North America. They have slowly recovered but are still quite rare north of the lake. Severe weather also decimated northern

(*Opposite*) The yellow warbler was much less common in John James Audubon's time because the brushy habitat it prefers scarcely existed in the old eastern deciduous forests. Like the common yellowthroat, this "yellow bird," as some call it, is a flexible nester, building in moist woods, old fields, gardens, and streamside willow trees. It often buries cowbird eggs in the bottom of its nest where they cool and fail to hatch. Yellow warblers are very common nesters in this area.

Blue-winged warblers like this one are closely related to golden-winged warblers, although they don't closely resemble each other. Because the blue-winged male is more aggressive than his golden-winged counterpart, he often successfully mates with golden-winged females. The resulting dominant and recessive hybrids are called Brewster's and (much rarer) Lawrence's warblers. This disruption of the mating cycle has enabled blue-wingeds to invade from the southwest. Although golden-wingeds are still well established in West Virginia, their cousins are replacing them farther north.

Bobolinks' buoyant songs can be heard in old hayfields in the Lake Erie region, though not so often as they were a century ago. These "blackbirds in formal wear" are declining around the lake like all grassland species. Besides losing open-country habitat, ground nesters like them are hurt by farmers' practice of mowing too early in the season before young have fledged the nest.

bobwhites, whose clear whistles pervaded my semi-rural childhood in northwestern Ohio. The severe weather of the late 1970s killed 90 percent of them south of the lake. Clean farming also limits bobwhites in heavily farmed areas. Although quail, too, have recovered somewhat during the recent mild decades, they are at their northern limits here, and hard winters will always give them trouble, as they do Carolina wrens.

The relationship between climate and other factors that influence nesting bird populations is a complicated one. For example, Harold Mayfield asserts in "Changes in the Natural History of the Toledo Region" that we can explain extensions and contractions of ranges by "the drying of the soil, the opening of the forests, the increase in brushy habitats, and the introduction of food-bearing plants"(48) rather than citing climatic warming over the past two centuries. But many observers credit a warming trend for southern birds' tendency to move north, whether or not that trend is a short-term one or a product of global warming. Roger Tory Peterson believed that for some birds, like cardinals, backyard feeding stations have helped support the push northward. In the case of other birds, such as cerulean warblers, hooded warblers, and Louisiana waterthrushes, we can see northern range expansions side by side with declines (such as in Michigan) because mature forest habitat has dwindled.

BIRDS ADAPTED TO HUMANKIND

In various ways, many species have adapted to a world transformed by humans. The most successful are often generalists that are flexible nesters and foragers. Cardinals like thick shrubby thickets and inhabit ornamental plantings in yards, parks, cemeteries, and other urban settings, as well as brushy sites throughout the countryside. Their adaptability is obviously one cause of their recent increases around Lake Erie. Gray catbirds, chipping sparrows, house finches, and robins are also fond of ornamental shrubbery. Many more robins sing at dusk in modern suburbs than did in the old forests; lawns and gardens suit their foraging methods for worms and insects during the nesting season. Unfortunately, this is not true for most birds.

A variety of species accepts the hospitality of birdhouses. Purple martins, house wrens, bluebirds and, unfortunately, aggressive house sparrows and starlings have adopted bird houses for lack of sufficient cavity nesting spaces in suburbs, old fields, and young forests. Martins have become dependent on their large "apartment houses," which are especially visible in Amish country. Tree swallows have increased in our area over the past few decades because they have begun using bluebird houses to supplement their own dwindling nest sites in tree

Adapted to shrubby thickets and hedgerows, gray catbirds are common nesters here. Though not quite such accomplished songsters as their cousins the mockingbirds, these birds' conversational songs are pleasant to hear, and their catlike calls identify them definitively. They produce two broods of blue-green eggs and deal with cowbird eggs firmly by ousting them from their nests.

cavities. Many common nighthawks have become city birds by nesting on the flat, gravel roof surfaces of buildings. Barn swallows stick their mud nests under eaves and in barns, and eastern phoebes have abandoned niches in cliffs or banks for bridges and other human structures. Few chimney swifts attach nests to the insides of hollow trees these days, and barn owls did not always nest in barns.

Even certain raptors now live close to human beings. Recent projects to restore DDT-decimated peregrine falcons to eastern North America have focused on city centers: These impressive birds will accept ledges on the twentieth floor almost as easily as on cliff sides and are free of great horned owls' attacks in the city. Then, too, the European rock doves that have now become city pigeons supply an all-too-abundant source of food for peregrine nestlings. American crows have found city streets and highways excellent sources of carrion, and their numbers are flourishing. Red-tailed hawks and kestrels perch near grassy interstate highway verges, good habitat for small rodents.

Cooper's hawks, too, have moved into urban environments, exploiting birds that feed or nest in parks and cemeteries.

Lake Erie is a central factor in area nesters' lives. Since large bodies of water heat and cool more slowly than the surrounding land does, the lake's effect is to moderate the region's climate—cooling down springs and summers, warming and extending autumns into relatively mild winters. The effects of the lake, combined with the fairly low land surrounding most of Lake Erie, extend the Eastern deciduous forests, in which so many of our nesters breed, northward.

We humans, however, are the ones who have transformed the terms on which birds breed and nest around the great lake. We have cut the forests, passing the advantage from birds that nested and foraged in their cool gloom to birds of the sunny fields and

At home nesting on a ledge in downtown Cleveland's Terminal Tower, this peregrine falcon finds ample food in the city's abundant rock dove population, as well as protection from other predators like great horned owls. Banning of DDT and a vigorous reintroduction program, especially in cities, have helped this superb bird of prey come back from the edge.

grasslands, the brushy thickets of abandoned farm fields, and recovering second- and third-growth woodlands. We are altering the climate through burning fossil fuels in ways that we ourselves don't yet understand, apparently favoring invaders from warmer southern climes. Our highways, suburbs, and shopping strips will, if sprawl is not slowed, gobble up more and more nesting and feeding habitat. Shoreline development has contributed to the well-documented decline of the rare piping plover, a small, attractive shorebird that will probably never nest again on Lake Erie and now is rarely seen even during migration. In fact, it is on the verge of extirpation in the entire Great Lakes region.

Finally, the states south of Lake Erie, as well as southern Ontario, have drained most of their original marshes, swamps, and other wetlands. Of the fifty U.S. states, Ohio is second only to California in the percentage of wetlands drained: more than 90 percent. Such habitat is more and more precious for supporting wetland birds such as herons—especially bitterns—ducks, rails, and others.

NESTING BEHAVIOR

One afternoon as I walked out of a café carrying a copy of Roger Pasquier's fine book *Watching Birds,* a young hostess followed me toward the door. "Are you one of those big bird fans?" she asked. When I said that I was, she asked with an intense expression if I could tell her where birds went to go to sleep. It took me a little while to realize that what she wanted to know was whether birds slept in their nests every night.

My inquirer, like many people who are not birders, was thinking of birds' nests as their homes, a notion encouraged by "birds in their little nests agree" and other sentimental phrases that float through our culture. After thinking for a moment, I told her that nests are really more like baby cribs than homes, that they are spaces for brooding eggs and enclosing helpless young rather than homes for avian families. Adult perching birds generally roost in protected places such as vine tangles and the depths of evergreens, grasping twigs automatically with tendons activated by the bird's weight.

Great horned owls are well adapted to a landscape of fields and woodlots; they have largely replaced barred owls, deep-woods birds that have declined in a landscape altered by humans. Great horned owls' varied diet includes occasional domestic cats or skunks, whose weapons do not appear to bother these powerful hunters.

Of course, there are exceptions. Tens of eastern bluebirds huddle in tree cavities to save body heat during cold winter nights, for example. Other hole nesters also may use cavities for nighttime shelter. I also explained that for many birds, nesting and foraging territories, which males and sometimes females defend vigorously from others of their own species, are more important as "homes" than the nests themselves.

Birds have evolved many specialized nesting and hatchling-care behaviors, all designed to ensure that new generations replace them. These variations are far too complex to cover adequately here, as are the varieties of monogamy, polygamy, and

One of the last birds to begin nest building, the American goldfinch prefers thistledown for lining its well-made nest and must wait for it to mature in summer. Often hidden in berry bushes, large egg clutches (five or six) keep goldfinches' numbers high. Homeowners who fill bird feeders with niger seed can attract these pretty and animated nesters well within the city limits.

polygyny displayed by different breeding birds. To generalize, however, birds have fashioned two main reproductive strategies. The first depends on female birds being able to feed well before laying relatively large, nutrient-packed eggs that hatch into well-developed young. These alert little fluffballs are called precocial young, a word related to "precocious." They can be up and out of the nest within hours, in some cases, and generally within two days, and they can forage for themselves. They also can maintain their body heat fairly early and require minimal brooding. Good examples of this type are shorebirds and waterfowl, such as ducks and geese.

The other strategy involves laying less-well-stocked eggs that hatch into naked, helpless altricial babies (meaning "requiring care") that remain in the nest for much longer before fledging. Because fewer energy resources go into the eggs, however, the hatchlings must garner a great deal of parental brooding, feeding, and other nurturing before striking forth from the nest.

The protein-rich insect diet their parents feed them helps them develop larger brains in relation to body weight than do precocial species. Thrushes, vireos, warblers, and many other songbirds have adopted this strategy, and parents have strongly ingrained feeding and brood-tending behaviors in these groups.

Many ground-nesting birds produce precocial young, especially those of the open country. After all, if babies can leave the nest quickly and scatter at a parent's alarm call, the brood as a whole has a better chance of escaping predators, even though single individuals may be picked off. Ground nesters also depend on speckled and otherwise camouflaged eggs to protect their embryos and may lay these in a mere scrape in the ground. Parents like killdeers may evolve broken-wing and other displays to draw prowling animals away from eggs or young. Though some ground-nesting wood warblers like ovenbirds hatch altricial young, they depend on building a well-camouflaged nest on the woodland floor to protect their eggs and nestlings.

Songbirds that nest higher off the ground in weeds, shrubs, or trees or on ledges and cliffs generally build more elaborate nests of great variety, though the nests of some raptors, cuckoos, and doves can be very sloppily made. Parents of altricial hatchlings must depend on nest camouflage and their own defenses against predators that are all too eager for a meal of fresh—or not so fresh—eggs or nestlings. These birds' eggs are also colored or camouflaged, except for those of hole nesters, whose eggs are uniformly white. These do not need camouflage because they are hidden from sight within the cavity.

Ehrlich, Dobkin, and Wheye's *The Birder's Handbook* is an excellent, nontechnical guide to the complexities of nesting and other aspects of bird behavior. The reader will find more fascinating detail there and in other books on behavior.

ADDITIONAL READING

Austen, Madeline J. W., Michael D. Cadman, and Ross D. James. *Ontario Birds at Risk: Status and Conservation Needs*. Don Mills and Port Rowan: Federation of Ontario Naturalists and Long Point Bird Observatory, 1994.

Beardslee, Clark S. and Harold D. Mitchell. *Birds of the Niagara Frontier Region*. Buffalo: Buffalo Society of Natural Sciences, 1965.

Brauning, Daniel W., ed. *Atlas of Breeding Birds in Pennsylvania*. Pittsburgh: Univ. of Pittsburgh Press, 1992.

Brewer, Richard, et. al., eds. *The Atlas of Breeding Birds of Michigan*. East Lansing: Michigan State Univ. Press, 1991.

Campbell, Lou. *Birds of the Toledo Area*. Toledo: *The Blade*, 1968.

Ehrlich, Paul R., David S. Dobkin, and Darryl Wheye. *The Birder's Handbook: A Field Guide to the Natural History of North American Birds*. New York: Simon and Schuster, 1988.

Fisher, Chris. *Ontario Birds*. Edmonton: Lone Pine, 1996.

Gingras, Pierre. *The Secret Lives of Birds*. Willowdale, Ontario: Firefly, 1995.

Levine, Emanuel, ed. *Bull's Birds of New York State*. Ithaca: Cornell Univ. Press, 1998.

Mayfield, Harold. "Changes in the Natural History of the Toledo Area Since the Coming of the White Man." *The Jack Pine Warbler* 40 (June 1962): 36–52.

Pasquier, Roger F. *Watching Birds: An Introduction to Ornithology.* Boston: Houghton Mifflin, 1977.

Peterjohn, Bruce G. *The Birds of Ohio*. Bloomington: Indiana Univ. Press, 1989.

Peterjohn, Bruce G., and Daniel L. Rice. *The Ohio Breeding Bird Atlas*. Columbus: Ohio Department of Natural Resources, 1991.

Theberge, John B., ed. *Legacy: The Natural History of Ontario*. Toronto: McClelland and Stewart, 1989.

Lake Erie Marshes

Why preserve the marshes? For our continued well-being on this planet, for much-needed recreational areas, for our aesthetic enrichment, for the benefit of those who come after us—in short, to hold on to the irreplaceable.

—Louis W. Campbell, *The Marshes of Southwestern Lake Erie*

Etienne Brulé landed at a river's mouth in what is now Ottawa County, Ohio, on November 1, 1615, and he named the stream Toussaint in honor of All Saints Day. As their canoes skirted the gray western Lake Erie shoreline, Brulé and his fellow Frenchmen must have witnessed clouds of waterfowl beyond modern imagination. Early November was the time of massive fall migrations. Geese and ducks spanned the continent between northern tundra,

Reduced to fewer than 1,500 pairs outside Alaska in 1982, bald eagles have rebounded in the last decade thanks to banning of DDT and strenuous efforts by wildlife managers. The Erie marshes provide vital habitat for these big birds, and sightings of eagles at the nest, quartering the marshes, or perching on muskrat houses are now frequent. Preserving adequate habitat is the most important factor in their conservation.

(*Opposite*) A great blue heron in breeding plumage stands motionless above its own reflection. This is the common marsh heron, and it can usually be seen peering intently into pools or flapping gravely over the cattail tops. Its diet is mostly fish, but it occasionally takes small mammals as well.

A pintail drake preens to keep his bridal attire in trim. Good numbers of these dabbling ducks appear in the marshes each spring. The species has declined sharply in recent years for unknown reasons. It is to be hoped that the handsome ducks will begin to recover in the new century as wood ducks did in the last.

prairie marshes, and wintering grounds on the Atlantic and Gulf Coasts. They gathered in the Lake Erie marshes to rest and feed in enormous gabbling, honking masses. "Goose music," as conservationist Aldo Leopold has called it, rang wildly along the cattail- and wild rice–fringed borders of the lonely lake.

The marshes formed a vast and busy world. From what is now the town of Vermilion to the mouth of the Detroit River in Michigan stretched an estimated 300,000 acres of marshes and swamp forest, covering an almost-flat area once occupied by a larger postglacial lake. Mile after mile of waving marsh vegetation harbored ducks and geese; herons, bitterns and rails; dabbling coots and diving grebes. The marshes also sheltered muskrat, mink, raccoons, foxes, weasels, and other furbearers and offered spawning grounds to many Lake Erie fish. Frogs and toads peeped and trilled in the cattails, and snapping turtles lurked in wait for the spring hatch of fuzzy ducklings. Bald eagles, northern harriers, short-eared owls, and other birds of prey hunted the sedge meadows and pools. In Ohio the marshes

Great blues build large groups of untidy nests high in trees. Sizeable heronries exist at Winous Point on Sandusky Bay, on West Sister Island, and at other places around Lake Erie. From West Sister adults fly to the marshes to forage for themselves and their nestlings. Created a National Wildlife Area in 1938, West Sister was designated a wilderness area in 1975 to protect its rookeries.

averaged about two miles wide and extended to the edge of the forbidding Great Black Swamp forest that in turn stretched far westward to where Fort Wayne, Indiana, now stands.

Brulé and the voyageurs paddled on, leaving the marshes to Indians, who trapped, fished, and hunted there, stuffing duck skins with leaves to use as decoys for waterfowl. Later, French settlers moved in. They lived not unlike the Native Americans—building bark huts, hunting, fishing, trapping, and trading—and they cleared little land. The nineteenth century, however, brought German immigrants whose minds were set on farming. To these people, a marsh was neither living space nor hunting ground, but a mosquito-infested wasteland that needed to be cleared and drained for productive use. With them began the demise of more than 90 percent of these marshes.

As early as the 1820s, another group focused on the marsh system's teeming wildlife. Sport and market hunters began to jockey for hunting rights to the better waterfowl areas, and the first duck club was founded at Winous Point in 1856. From then on, farmers and hunters coexisted in the area. By 1900, farmers had drained and diked large sections of what they called the "pumplands," but many found marsh cultivation both expensive and risky. Northeast winds could pile up seiches (pronounced "sayshes"), windblown tides that might breach dikes and drown crops in a few hours. Lake levels could also fluctuate widely over the years, as lakefront condo owners have more recently discovered to their dismay. Many farmers gave up and let the pumplands reflood.

Meanwhile, market hunters and rich city sportsmen alike exploited ducks and geese for decades. Market hunters would shoot, clean, salt, and pack ducks by the barrel for Chicago and New York markets. Those familiar with Edith Wharton's novels may recall that canvasback ducks were especially prized for high-class New York banquets. During the 1870s wild ducks sold for about ten cents each. From 1880 to 1903, Magee Marsh, now a state wildlife refuge west of Port Clinton, Ohio, was known as the Crane Creek Shooting Club. Sportsmen traveled to it from as far as Pittsburgh, Cleveland, and Detroit for the hunting.

Droves of common grackles migrate throughout the area, among the earliest to arrive in spring. These aggressive birds are very common. They are opportunists, eating invertebrates, small vertebrates, bird eggs and nestlings, fruits, nuts, and grain. Their nest-robbing propensities are blamed in part for recent songbird declines.

An airborne great egret is one of the most beautiful sights the marshes have to offer. Egrets' breeding plumes, or aigrettes, were once so popular for ladies' hats that plume hunters pushed the birds close to extinction. As hunting pressures mounted in the south, egrets and black-crowned night herons extended their ranges farther north. They are at the northern edge of their range in this region.

The duck hunters, like the farmers, began to dike the marshes to control water levels, but they diked and pumped water not off the land but onto it, to attract both muskrats and waterfowl. In 1920 muskrat pelts brought four dollars apiece and helped defray diking expenses. From the late 1920s to 1950, Magee Marsh was leased to a private duck club that annually took 2,500 ducks from the cattails, mud flats, and reedbeds of the marsh. From Maumee Bay on the west, along the Lucas and Ottawa County, Ohio, shorelines, and around the big embayment of Sandusky Bay to the east, hunting clubs controlled thousands of acres of prime marshland, a fact that explains the survival of considerable tracts.

"Considerable" is a relative term, of course. Of the voyageurs' 1,500 square miles (4,000 km) of wild wetland, only about forty (100 km) remain now around Erie's western basin.

(Wetlands are much scarcer on the eastern basin because that part of the lake is so much deeper.) Some of the marshes still belong to shooting clubs, but the U.S. Fish and Wildlife Service, Canadian Wildlife Service, and Ohio and Michigan state and local governments administer others.

Times have changed since only hunters, fishers, and trappers would spend time in a marsh. Until recently, most people thought of wetlands as the early farmers did—as wastelands or good places to dump trash, empty except for clouds of blood-sucking insects. Sheldon Marsh on Sandusky Bay, bought and protected in the early 1950s by Dr. Dean Sheldon, was dubbed "Sheldon's Folly" until 1979, when Ohio dedicated its excellent lake, forest, swamp, and marsh habitats as a nature preserve.

While all marshes are wetlands, not all wetlands are marshes. Ecologists define marshes as saltwater, brackish, or freshwater wetlands characterized by emergents, soft-stemmed herbaceous plants like cattails and pickerelweed that stand with their roots in the water and their stems above it. These marshes are termed

Large numbers of blue-winged teal, among the smallest of the puddle ducks, visit the marshes in April, and some remain to breed. These attractive males show the distinctive white facial crescent, absent in blue-winged females. Northern shovelers arrive at approximately the same time.

"bittern-rail cattail marshes" in our area. Unlike northern swamp forests and bogs, which host relatively few life forms, marshes teem with myriads of plants, birds, mammals, amphibians, fish, invertebrates, and microorganisms of all kinds. If bogs, acid, still, and deep, are what the Irish poet Seamus Heaney calls "the memory of a landscape," then marshes must represent the flux of its present moment, seething with life, motion, and drama. Ground water feeds them, and lake waves and ocean tides replenish oxygen for many of them, which helps explain their amazing productivity.

An Erie marsh is a place whose moods change according to weather and season. Chill spring winds tease the water into waves and sweep rafts of diving ducks in and out of its cold gray troughs. The breeze ruffles dry rushes, sedges, and cattails and whips the comical topknots of displaying mergansers, or "fish ducks." Mallards, shovelers, and pintails circle the marsh in courtship flights, landing with much ado, along with many other species of diving duck, such as ring-neckeds, redheads, lesser scaup, and buffleheads. The vanguard of elegant tundra swans

Lake Erie's marshes include many habitats—swamp forest, sedge meadow, and deep-water marsh—but the most important portions are bittern-rail cattail marsh, in birders' terminology. Here an American bittern treads through the cattails at Conneaut Marsh in northwestern Pennsylvania. Loss and degradation of marsh habitat account for decreases in populations all across its range.

Mallards are the most abundant marsh waterfowl and nest commonly there. Since the 1940s this handsome and, as hunters will affirm, tasty species has increased dramatically in other parts of the lake region as well. Competition from mallards has contributed to declines of the closely related American black duck. The two species often hybridize.

also arrives following the retreating ice northward. Agitated geese gabble, flashy red-winged blackbirds and pied-billed grebes call, and killdeer pipe in the distance.

High above, red-shouldered hawks ride thermal air currents north. Herons and egrets, already in breeding plumage, hunt in the shallows. Nature pulses with color, sound, movement, and life. Later in spring, waves of brightly colored songbirds rest and feed in trees near the water waiting for favorable winds before attempting the Lake Erie crossing. May is a good time to look for shorebirds. Look through the legions of dunlins to find rarer species: whimbrels, Hudsonian godwits, phalaropes, and the small sandpipers collectively called "peeps." In June the pace

Birdwatchers can see small numbers of solitary sandpipers in both spring and fall as these shorebirds migrate to and from their nesting grounds in boreal forests and muskegs. Stirring up the water with its leading foot, this sandpiper feeds on a variety of land and water insects, worms, spiders, mollusks, crustaceans, and frogs.

(*Opposite*) Canada geese begin nesting earlier in spring than most birds do, often choosing the top of a muskrat lodge for their nursery. This location may offer some protection from predators. By the end of April, flotillas of downy goslings appear attended by both parents. Unlike most drakes, Canada ganders are attentive spouses and fathers.

slows, and the marshes quiet down, a good time to see mother mallards and ducklings or Canada geese with gaggles of young.

At dusk on a July evening, little seems to move. Heat presses like a moist palm on the marsh's surface. Birds are quiet except for swallows hawking low above the water after insects. Black ducks and male mallards in eclipse plumage skulk quietly in the vegetation, temporarily grounded to moult last season's flight feathers. The white forms of great egrets hang against the horizon like elongated s-hooks, as they and their relatives, the great blue herons, scan the waters for small fish and amphibians. Green herons and rails stand motionless among the lush green of cattails and other marsh plants or slip invisibly through their deep shadow. Turtles, snakes, and frogs hide in the cool mud.

This is the time for showy water plants. Swamp rose-mallows' pink lampshades glow above the still water, which also mirrors the blue spires of pickerelweed, white arrowhead spikes, purple loosestrife, and masses of yellow bur marigolds. A muskrat glides slowly, its head cutting a v-shaped wake among the floating leaves and fragrant blooms of water-lilies, spatterdock, and the beautiful American lotus, which unfurls eight-inch yellow flowers and leaves up to two feet across.

Bustle and sound return again at summer's end with small flocks of shorebirds moving south. October brings masses of coots, American wigeon, and blue- and green-winged teal, as well as many other ducks. In November fifty or more snowy tundra swans may float and feed serenely on the water. The main Canada goose migration sweeps out of the north during this month. Because the Lake Erie marshes lie at a crossroads of the Atlantic and Mississippi flyways, and because birds pause to rest there before and after the Lake Erie crossing, migrations of many kinds of birds—waterfowl, warblers and other songbirds, shorebirds, and birds of prey—are often spectacular. This is especially true on a day after a strong north wind, though migration's pace is generally slower than in spring.

When the geese have moved on, leaving several thousand resident birds behind, the marshes subside into winter stillness again. Although the land may look deserted, muskrats forage under water, and mink, weasels, eagles, a few rough-legged hawks from the north, harriers, red-tails, American kestrels, and short-eared owls hunt for wary rodents. Song and tree sparrows forage like mice among the winter foliage along with dark-eyed juncos and American goldfinches in pale winter plumage.

(*Overleaf*) Nesting in moist woods and occasionally in wet meadows, American woodcocks are nocturnal and solitary. The male is most easily seen in spring when he performs a virtuoso display flight to impress his intended. These chunky sandpipers are fairly common in proper habitat around Lake Erie and throughout the eastern United States and southeastern Canada.

Beneath the ice many of the tiny organisms that form the complex food webs of the marsh remain active or hibernate until the spring thaws begin.

Of course, these marshes have changed over the years. Diking by duck clubs stabilized water levels and increased waterfowl and other wildlife, especially those birds that prefer the edges of water and vegetation. The dikes provided upland nesting sites for ducks such as blue-winged teal and for mammals and turtles as well. But they also encouraged raccoons, skunks, opossums, and snapping turtles, all of which savor eggs and nestlings, and they discouraged lake fish from swimming into marshes to spawn. When carp were introduced into these diked areas during the late nineteenth century, the marshes' vegetation also changed. Carp caused irreparable damage, stirring up the bottom silt and eliminating valuable food plants, like wild rice, that required clear water to thrive. The activity of these alien fish also helped aggressive plants to proliferate, such as giant reed grass

It seems hard to believe that the exotic colors of this male wood duck belong in the temperate zone. However, the species is fairly common here in woods near water, nesting in tree cavities or in elevated houses provided by wildlife managers and others. The birds were heavily hunted and declining fast until the last half-century, during which hunting restrictions and recovery efforts enabled populations to rebound.

(phragmites) and purple loosestrife, which offer little value for wildlife. Muskrat farms have also devastated marshes, with the animals literally eating themselves out of the cattails. These "eat-outs," as they are called, have been another factor in the decline of quality Erie marshes.

Today, smartweed, millet, and rice cutgrass have replaced earlier food species, attracting mallards, American black ducks, blue-winged teal, and other dabblers rather than the diving ducks that once nested in the marshes. Wood ducks, mallards, blue-winged teal, and a few black ducks are the most common breeders, with small numbers of hooded mergansers, northern shovelers, pintails, redheads, and ruddy ducks. Divers like common mergansers and ring-necked ducks now use the area as a stopover on their way farther north or south rather than as nesting territory. Management for waterfowl keeps water levels artificially high, limiting the mud flats shorebirds' need for foraging. The above problems are especially true of Metzger Marsh, another state-managed wetland in northern Ohio west of Magee Marsh.

Like wood ducks, small numbers of hooded mergansers nest in tree holes in wooded swamps around Lake Erie. The male's striking white patch shows clearly when he raises his crest, as this one is doing. Narrow, serrated bills help mergansers to capture slippery fish, which are their main food, along with other small aquatic creatures.

Relentless agricultural, population, and development pressures threaten the Lake Erie marshes, as they do most natural areas in the United States. Agriculture alone has destroyed more than three-quarters of the United States' wetlands over the past two hundred years. On Ohio's lakeshore, marinas and condos spring up like mushrooms, preempting farmland and wetland alike, causing land prices to soar, and making the acquisition of natural areas difficult. Changes in drainage patterns caused by building and increased silt loads from farm runoff and development also damage wetlands. The past half-century has seen disastrous pollution from DDT and other pesticides, agricultural fertilizers, and industrial PCBs.

Nevertheless, the news is not all grim. Despite ecological changes and shrinkage of wetland area, the marshes at Lake Erie's western end still harbor about 300 bird species, of which nearly 150 are known nesters; an estimated 1,000 vascular plants; about twenty-two mammals, not counting bats; as well as reptiles, amphibians, fish, and invertebrates. Since midcentury, government has protected thousands of acres of prime marshland. In 1938 Franklin D. Roosevelt signed Ohio's first national wildlife refuge into being at West Sister Island in Erie's western basin to conserve its massive rookeries of great blue herons, black-crowned night herons, great egrets, and other wading birds. Recently, nesting double-breasted cormorants—virtually absent two decades ago—have exploded at West Sister. The island is off-limits to the public to protect vital breeding populations.

At various sites, wetland enhancements have gone forward, such as at Michigan's Pointe Mouillee and at Pickerel and Pipe Creeks near Cedar Point in Ohio. Diking and reclaiming marshes from the lake itself is often cheaper than trying to acquire land already above lake level. However, the quality of these created marshlands is usually not very good, with disturbed bottoms vulnerable to phragmites and loosetrife. Still, they are better than no marsh at all.

Ottawa National Wildlife Refuge joined West Sister in the early 1960s. There the U.S. Department of the Interior's Fish and Wildlife Service manages several units totaling about 8,500 acres. These include the main unit, midway between Toledo and Port Clinton, Ohio; West Sister; Darby Marsh, just west of Port

Among the most attractive of swallows, tree swallows are common cavity nesters in the Lake Erie region, using tree holes, hollow fence posts, and nest boxes. Starlings are a problem for these pretty birds, since they are also cavity nesters and are more aggressive than tree swallows and bluebirds. Tree swallows arrive earlier in spring and stay later in the fall than any other swallow, migrating in large flocks.

Here the camera catches a retiring least bittern, America's smallest heron. These birds are well camouflaged and freeze in their tracks to escape notice, holding their long necks vertically among the cattail stalks as do their larger cousins, American bitterns. Few who do not wade out into the marsh itself ever see least bitterns, and many consider them to be endangered. However, they are probably more common than most people think.

Clinton; and Cedar Point National Wildlife Refuge on Maumee Bay, east of Toledo.

Originally established as a rest stop for migrating waterfowl, Ottawa was charged in 1974 with protecting habitat for endangered species as well, especially migrating peregrine falcons, resident bald eagles, rare Blanding's turtles, and melanistic garter snakes. Refuge headquarters also manages wildlife on 591 acres of Navarre Marsh, owned jointly by Toledo Edison and the Cleveland Electric Illuminating Company. While Ottawa's main unit offers dawn-to-dusk hiking on eight miles of trails every day of the year, the service restricts entry to other units.

In 1951 private owners sold the land for Crane Creek State Park and the surrounding 2,600 acres of Magee Marsh Wildlife Area to the Ohio Department of Natural Resources because dike maintenance had become too costly. These lie two miles east of Ottawa Refuge's main unit. The ODNR Division of Wildlife also manages Metzger Marsh, a diked wetland west of the Ottawa refuge, as well as smaller wildlife areas at the Toussaint River,

Little Portage, Willow Point and Pipe Creek on Sandusky Bay, and at Resthaven near Castalia. The Division of Natural Areas and Preserves has charge of Sheldon Marsh, located on Sandusky Bay between Sandusky and Huron, and Old Woman Creek National Estuarine Sanctuary and State Nature Preserve just east of Huron. In Michigan, fine marshes at Lake Erie Metropark are managed by that state's Department of Natural Resources.

Citizens' groups have also become increasingly active in wetland preservation. An example is The Nature Conservancy's acquisition of a 2,100-acre marsh on Sandusky Bay in 1986. In

Female wood ducks are as attractive in their own modest way as their showy mates. Eggs and newly hatched ducklings are visible in the bottom of this natural tree cavity. When it is time for the tiny young to fledge the nest, the mother will coax them to jump as much as fifty feet to the ground. From there they will make their way in tandem to the nearest water.

this Pickerel Creek Wildlife Area grow rare prairie white-fringed orchids. Three pairs of bald eagles nested there in 1998.

No other marshes around the lake rivaled the great expanses on its western basin, and drainage and damming of streams have further diminished the region's wetlands. Conneaut Lake in northwestern Pennsylvania once had productive marshes that were ruined by early cottage and amusement-park development. Nearby, one of Pennsylvania's finest wetlands, Conneaut Marsh,

A male northern shoveler bursts from Medusa Marsh on Sandusky Bay, Ohio. This is one of several marshes in the bay area. Like other puddle ducks, shovellers can take off from water without pattering along its surface, as diving ducks must do. The large spatulate bill, longer than the head, is a distinctive field mark for males, as well as for the drabber females, and is the source of the species' common name. Shovelers feed mainly by sieving small aquatic plants and animals through their bills' comblike edges.

escaped the development that transformed Conneaut Lake and is a good place to look for black terns, bitterns, wood ducks, and others. As long ago as 1868, however, Pennsylvania passed an act to channelize and drain the marsh, and it remained a meadow until the 1920s. Eventually the channel filled in with soil eroded from surrounding farmlands, and ditches caused water once held in the fields to drain into it. Today the marsh is a valuable resource for both hunters and naturalists. About 80 percent of southern Ontario's wetlands has been severely altered or destroyed, and Ohio's 90 percent wetland loss statewide is in the United States second only to California's 91 percent. What marshes remain are precious habitat indeed, and more should be protected.

Southwestern Ontario contains some of the most outstanding marshes in the Lake Erie region, such as that at the St. Clair National Wildlife Area on Lake St. Clair. This is one of the very

Canada geese fly out of the marsh in spring. The sight of these birds is much more common than it was even twenty years ago. Like wood ducks, Canada geese were in steep decline at mid–twentieth century. Restocking efforts and hunting restrictions resulted in a population explosion of the giant subspecies in this area. Giants are largely nonmigratory if there is sufficient food, and many stay in the marshes all winter, breeding there in spring. Interior Canada geese nest to the north on James Bay and use Erie's marshes for resting and feeding stopovers during migration.

few nesting places for king rail and yellow-headed blackbirds in the area. At Walpole Island, where the St. Clair River enters that lake, there are superlative marshes, but they are Indian territory, and permission to enter them must be approved by the First Nation Chief and Council.

On the southern side of the Ontario peninsula along Lake Erie's north shore, Long Point, Rondeau Provincial Park, the Point Pelee area, and the Big Creek marshes at Holiday Beach Conservation Area near the Detroit River are most notable. The large marshes at the base of the Long Point World Biosphere Reserve, designated by UNESCO in 1986, are home to least bitterns, common snipe and sedge wrens. Forster's terns and little gulls also nest there, and waterfowl stage in huge concentrations during late March and early April. However, only the base of the long sand spit that forms Long Point is accessible, and much of the area is private property. Birders may view the extensive cattail marshes at Rondeau Provincial Park much more easily from the 15 kilometer (10 mile) round-trip walk on the marsh trail there. (Rondeau is located between Long Point on the east and Pelee on the west.)

The Pelee marsh, another high-quality site, used to be twice as large as it is now, forming a continuous wetland with Hillman Marsh to the northeast, as Tom Hince points out in *A Birder's Guide to Point Pelee.* Drained portions of the old wetland are often used to grow onions. These areas can provide good forage for shorebirds and gulls under the right conditions in a region very short of adequate shorebird habitat.

Bald eagles, along with now-abundant Canada geese, are a prime example of how changed public attitudes and improved management techniques can rescue wetland species once on the brink of oblivion. Scientific management has increased Canada geese from a mere 53,000 on the Mississippi flyway in 1946 to the following dramatic numbers at the end of the century: approximately 135,000 geese (interior subspecies) that nest on James Bay; 1,390,000 largely nonmigratory Canada geese of the giant subspecies; a tallgrass prairie population of 550,000; 425,000 geese on the Atlantic flyway; 970,000 in the Mississippi Valley; and 270,000 Eastern prairie (interior) Canada geese. An Ohio Division of Wildlife project boosted their populations dramatically in this area when in 1967 the division installed a Canada Goose Management Investigations branch at Magee Marsh. Workers released twenty wing-clipped pairs of geese that year. Those birds and wild ones attracted to them increased to 1,082 by 1972 and in that year raised 703 goslings. Canada geese are now common in the Lake Erie region—more common than some would like. Many of the giant subspecies overwinter and nest in the marshes in spring. At Pymatuning Lake near Lake Erie on the western Pennsylvania border, the Pennsylvania Game Commission reports peak fall populations of about 15,000 geese in the refuge at one time, as flocks arrive and depart, as well as about 3,000 resident geese. And 10,000 to 20,000 winter in the western Erie marshes.

Many fail to realize that the "first robin of spring" is often a bird that has stayed in the area all winter. It has simply withdrawn from suburban summer lawns to sheltered areas with adequate food supplies, largely berries and other fruit. The Erie marshes host great roosts of American robins during fall and winter months.

DDT poisoning decimated bald eagles during the 1960s and 1970s, thinning their eggshells so that the eggs shattered before embryos could develop. Though fifteen nests overlooked the marshes in the late 1950s, the 1979 midwinter count found only six birds. Wildlife biologists at the Ohio Department of Natural Resources labored for a decade to bring the eagles back. They placed eaglets fostered by museums, zoos, and the Patuxent Wildlife Research Center in Maryland into nests for adoption and carefully monitored the eagles' progress.

For years the project must have seemed like a lost cause. In 1985 only three pairs of birds successfully raised their own young, though two foster chicks fledged the nests as well. The next two seasons, however, marked the turnaround. Nine eaglets fledged in 1986, eight of them native birds. In March 1987 twelve pairs nested, and growing reports of migrant eagle sightings showed that the great birds were beginning to recover in North America, due both to recovery efforts and to a gradual drop of DDT levels in the environment. In 1999, Ohio boasted fifty-seven nests, many near Lake Erie, and seventy-three eaglets successfully fledged. Birders' reports of dramatic eagle sightings soared. Tom Hince reports that in 1999 five active nests existed in the immediate region of Ontario around Point Pelee. Two are known in Michigan between Detroit and the Ohio line. In Pennsylvania there were only three bald eagle nests between 1963 and 1980; in 1996 the nesting population had climbed to twenty pairs, eight in the Pymatuning area.

For the geese, and more especially for the eagles, adequate habitat is now the crucial problem. Luckily, Lake Erie eagles have adapted to living near human beings. Eaglets see tractors, cars, and people from their nests, and they have come to accept this situation as normal. The eagles also have adapted to building their nests in small woodlots, so hunting territory is now more critical than nesting sites. Large raptors need a lot of land, and it's not clear yet how many nests the diminished marshes can still support.

Lake Erie's marshes offer a taste of wilderness, despite their history of human disturbance and their location near urban areas, and they provide crucial nesting and migrating habitat for many birds. Watching far-off masses of bay ducks staging and pairing for the upcoming breeding season or common mergansers and ring-necked ducks working their way through marsh vegetation on a clear spring day helps one appreciate distance, space, and wildness. To see skeins of Canada geese unraveling against an iron-gray sky and to hear their music or the whistling rush of a tundra swan's pinions almost takes one back to that All Saints Day of 1615 when French canoes slid out of the lake mist into the Toussaint River's mouth.

It's a great piece of luck that a rich slice of the Erie marsh ecosystems has survived human exploitation, unlike Illinois's Kankakee marshes or Ohio's Great Black Swamp, both of which evaporated from the landscape, leaving few traces. With continuing luck, awareness, and vigorous effort, we and our children may still enjoy the elegant silhouette of a great blue heron against the water, hear the racket of a saucy marsh wren or the "witchety witchety" of a common yellowthroat far off in the cattails, and watch with awe a bald eagle land at a nest to feed its young.

ADDITIONAL READING

Bolsenga, Stanley J., and Charles E. Herdendorf. *Lake Erie and Lake St. Clair Handbook*. Detroit: Wayne State Univ. Press, 1993.

Campbell, Louis W. *The Marshes of Southwestern Lake Erie*. Athens: Ohio Univ. Press, 1995.

Eckert, Allan W. *The Wading Birds of North America (North of Mexico)*. Garden City, N.Y.: Doubleday, 1981.

Grimm, William C. *Birds of the Pymatuning Region*. Harrisburg: Pennsylvania Game Commission, 1952.

Hince, Tom. *A Birder's Guide to Point Pelee and Surrounding Region*. Wheatley, Ontario: Hince, 1999.

Niering, William A. *The Audubon Society Nature Guides: Wetlands*. New York: Knopf, 1985.

———. *The Life of the Marsh*. New York: McGraw-Hill, 1966.

The North Wind

In our enthusiasm to glorify and embrace nature, superlatives have lost their potency to describe its grandeur. We find ourselves repeating the same cliches. Those of us who write about bird migration are as guilty as anyone, but the facts of the matter are such that there is really no need for hyperbole. Twice each year, billions of birds, entire species, swarm across the globe, traveling thousands of miles as they follow the sun to populate regions that are habitable for only part of each year. The spatial scope of these migrations exceeds all other biological phenomena.

—Kenneth B. Able, *Gatherings of Angels*

In contrast with spring migration, the fall movement of birds in our area seems a quiet and leisurely affair. Springtime birds may sport colorful—even gaudy—breeding plumage, and their songs announce and animate each sunrise. In that season, the race is on to reach northern nesting grounds, to claim territories, to court, to mate, to nest, to fledge young, and to stage for the journey south again. Summer is almost too brief a season to achieve this program, especially for birds that nest on the arctic tundra or in the boreal forests of Canada.

Autumn individuals seem different creatures entirely: Most have exchanged their flashy mating garb for modestly hued traveling dress as they slip quietly southward. Their songs are mute, and we humans are often deaf to the subdued call notes that surround us in autumn. But massive movement is going on across the hemisphere and around the world. The entire migration covers a full six months, from the first shorebirds that appear during late June and early July to the owls, gulls, and winter finches of December. Actually, the scope of autumn migration is as great or greater than that of spring—though its patterns and timetables can be quite different. Throngs of newly fledged birds swell numbers vastly, and the season boasts its own unique moments of spectacle.

The fall vanguard is made up of shorebirds from the two main families, plovers and sandpipers. Birders distinguish these

(*Opposite*) Broad-winged hawks kettle above the hawk-watching tower at Holiday Beach, Ontario. The first north winds after September 10 trigger their spectacular migration. Lake Erie Metropark at the mouth of the Detroit River in Michigan tallies even larger numbers of these long-distance migrants. If a front is approaching from the west when a strong northeast wind is blowing, countless thousands of broad-wingeds may be funneled past these lookout points on the Canadian and American sides.

Greater yellowlegs, a species of sandpiper, sweep the water with their forcepslike bills to catch small fish. They are common spring and fall migrants around Lake Erie. This individual looks smug after a successful bout of fishing at Conneaut Harbor, Ohio. Yellowlegs' autumn movements may actually span five months! Ornithologists have pointed out that fall migration should really be called postbreeding migration to be completely accurate, since it begins in early summer and extends through into winter. Some birds begin moving north again as early as January, so prebreeding migration is a more accurate term than spring migration, as well.

by their feeding habits. Plovers are "pickers" and have large eyes and forage by sight; they scurry along the ground, stopping to peck when they locate prey. Watch a killdeer in an empty lot to see the pattern. Sandpipers tend to be "probers," hunting by feel in the mud or sand with their longer, sensitive bills. As they explore the burrows of small invertebrates, their heads have been described as mimicking the bobbing action of sewing machines.

Most shorebirds don't linger to breed in this region, except for American woodcock, spotted sandpipers, killdeer, and, rarely, common snipe and upland sandpipers, which are grassland birds declining in the eastern part of their range. The many species that pass north over our area have evolved instead to exploit the burst of insect life fueled by long hours of Arctic sunlight. During the short but intense northern summers, they wade and feed in the shallow lakes of meltwater cupped above permafrost.

These migrators' major flights follow the Atlantic and Pacific Coasts or span the Great Plains west of the Mississippi. Several routes leap spectacularly over the ocean from the Canadian maritime provinces and New England all the way to South America. The Great Lakes region is relatively shorebird poor. Yet the official bird list for the region includes forty-seven species, more than for either warblers or waterfowl. Moreover, notes Bill Whan in the journal *The Ohio Cardinal*, that state "provides the largest expanse of potential stopover habitat in the eastern US between the Atlantic coast and the breeding ranges of most of these migrants." He continues, "Hundreds of years ago the shore of Lake Erie, and to a lesser extent the wet prairies to the south, must have teemed with migrant shorebirds every spring and fall. That this no longer happens seems almost entirely the result of human ignorance in some cases, and human insouciance in the rest" (22:37).

Market and recreational hunters routinely slaughtered shorebirds before the federal government protected all but woodcock and snipe as nongame species early in the twentieth century. The fat little Eskimo curlew was shot to the brink of extinction and

Bonaparte's gulls are common in fall in the Lake Erie region, and observers can view thousands of staging birds in November and December. More are seen in that season than in spring because the pace of migration is slower in autumn. The birds are en route from their summer range in northern Canada and Alaska to the Atlantic and Gulf Coasts. This gull poses in postbreeding plumage and lacks the black head of the summer adult.

may, in fact, have disappeared entirely. Some other species have not fully recovered from an age when wagonloads of birds were shot and then dumped to rot in order to clear space for still more dead birds. In addition, many of the shorebirds that do ride the wind today must ride it past the Lake Erie region rather than stopping here. This is because suitable foraging habitat has nearly disappeared: Most of the area's wetlands have been drained.

Ironically, these birds' very status as nongame species has worked against them. Because shorebirds are no longer hunted, government agencies have traditionally managed what marshes remain not for them but for waterfowl, which hunters may bag legally. Hunters pursuing waterfowl—and as a group spending large amounts of money to do so—are usually not interested in preserving expanses of mud or beach. As Whan wryly comments, "We haven't worked out a way to assess the value of animals we can't take home with us" (85). Property owners view the flats as waste ground that can easily be rendered more profitable. Natural shorelines have also been diked to protect shorefront property and roads, eliminating still more feeding habitat. Some areas do provide good shorebird habitat, such as the limestone shelving along Erie's northeastern shore in Ontario. Rarities such as American oystercatchers and wandering tattlers occur when low water levels expose these rocky ledges.

Despite the problems, observation can reward birdwatchers who love the energy and subtlety of these wild brown birds. This is especially true in years when low water levels expose mud flats rich with invertebrate life or when westerly winds create such flats in the Erie Marshes. An arresting example of the importance of proper feeding habitat occurred in 1994 at the Turtle Creek unit of Magee Marsh Wildlife Area in Ohio. There, work on a diking project exposed a large mud flat submerged for decades, and thousands of shorebirds appeared suddenly to forage on it: 61,742 birds of thirty-two species that autumn and 34,331 of twenty-six species the following spring before the flat was reflooded. The numbers were striking indeed and included a surprising variety of rarer species. This example shows that if mud is provided, shorebirds will come.

Those who care about preserving whole natural systems rather than only showy species or those that can be hunted should encourage state and provincial managers to plan for

(*Opposite above*) This adult little gull belongs to a Eurasian species much sought after by North American birders. It now breeds irregularly on this continent from the Great Lakes to Hudson Bay, and a few birds winter in our area. The Niagara River Gorge is a great place to see this "specialty bird" of the Lake Erie region in autumn. If it leaves this area, it will winter on the Atlantic Coast.

(*Opposite below*) The buff-breasted sandpiper is a bird of dry, grassy fields and dry mud flats. It is an accidental spring migrant and uncommon in fall, a season when juveniles are consistently observed. Most adults migrate through the eastern Great Plains. This is another specialty bird of our area, and to see eight to ten in a season is cause for self-congratulation.

shorebirds as well as for ducks. Unfortunately, it appears that few managers think about moving in that direction at present. In fact, Metzger Marsh, one of Lake Erie's most significant stopping and resting areas for shorebirds in Ohio, was recently rediked to reclaim wetlands—or so it was said. As a result, invasive species such as phragmites and canary grass have invaded the marsh creating an environment undesirable for bird life. One questions the logic of this decision.

Diminutive northern saw-whet owls nest in the boreal and transition forests of Canada, New England, and the Appalachians. More northerly populations shift southward in autumn, though they are easily overlooked on their winter range; saw-whets are uncommon spring and fall migrants, and a few are probably residents. This owl was located at Mentor Headlands on Lake Erie east of Cleveland, Ohio. The Headlands is a small preserve—only about 100 acres—but over 300 species of birds have been recorded there.

By early July, greater and lesser yellowlegs, among the many sandpipers that nest in northern muskeg and tundra, have begun to arrive here. With them come least sandpipers and gangs of short-billed dowitchers, soon followed by semipalmated plovers, sanderlings, and a number of other sandpipers (semipalmated, upland, solitary, pectoral, and stilt). By the end of the month, birders can hope for black-bellied plovers, ruddy turnstones, and a few Wilson's phalaropes. The first landbirds appear as well, including flycatchers (least and yellow-bellied followed by alder in early August), Swainson's thrush, Tennessee warbler, ovenbird, and northern waterthrush. Banding evidence suggests that orioles and yellow warblers may actually begin to move south as early as the first week of July.

Why do shorebirds leave the northland so early in the year, when, by all accounts, ample food is still available there? The answer is not completely certain. In general, females leave the

Common terns no longer breed in this area in any numbers. Loss of beach habitat and harassment by predators such as mammals and the expanding ring-billed gull population have all but eliminated the species as an Erie nester. Experiments with artificial breeding platforms have met with some success, however.

Another sought-after bird in this region is the purple sandpiper. It is a rare but regular migrant in one's and two's. Birders search for it on stone jetties and natural rock outcroppings around Lake Erie. Purple sandpipers' bills are curved slightly downward; shorebirds can often be identified by the differing lengths and curvatures of their beaks, which have evolved to suit varied foraging habitats and methods.

breeding grounds first and arrive at staging areas on the coasts and at inland sites the earliest. They have expended the most energy in producing eggs and must lay on fat before leaving the Northern Hemisphere behind. Apparently, time spent fattening on the staging grounds is key to a successful migration. Males follow, trailed last of all by the juveniles, which make up a large percentage of migrants seen in our area. These young ones' innate sense of direction serves to guide them south without any help from their elders, and they may not even follow the same routes their parents do.

Shorebird migration continues in flood through August, a month when American golden plovers and buff-breasted sandpipers appear, joined by succeeding waves of landbirds. Many of these are warblers, challenging to identify in their postbreeding plumage: black-and-white, magnolia, Cape May, blackburnian, chestnut-sided, bay-breasted, mourning, Wilson's, Canada, and by the end of the month blackpoll, Nashville, black-throated blue, yellow-rumped, and black-throated green. Other neotropical migrants include scarlet tanagers and rose-breasted grosbeaks, and Baltimore and orchard orioles.

Purple martins, bank swallows, and barn swallows flock on wires in September as they begin the journey to Central and South America, with the majority of the purple martins gone by the middle of the month. This is also the time to look for migrating flocks of nighthawks in the evenings. Tree swallows will follow later, their numbers peaking in October. Hunters of aerial insects generally migrate by day—unlike most songbirds—and eat as they move. Eastern kingbirds, too, are seen in large numbers by the end of the month. Flights of Bonaparte's gulls and common terns are reported at Point Pelee during August, though both birds have been noted in small numbers throughout the summer.

A juvenile sanderling demonstrates preening, which birds must do constantly to keep their feathers in good shape. This is essential both for flight and for waterproofing. Like many other shorebirds, most sanderlings seen in the fall are juveniles. They chase the waves on beaches to snatch up small invertebrates, and their comical gait has been compared to that of a small clockwork toy. Like many other shorebirds, sanderlings are declining in North America; changes in Arctic weather patterns may be the culprit.

Songbirds continue to arrive in September, including white-throated and Lincoln's sparrows, palm warblers, winter wrens, ruby- and golden-crowned kinglets, blue-headed vireos, hermit thrushes, and American pipits. A strong north wind can bring a rush of hummingbirds. However, most of these birds will be gone by the end of the month except for winter residents. Insect eaters particularly must move away to the south before chilly weather eliminates their food supplies. Those that remain, such as juncos and chickadees, are either seed and fruit eaters or species that forage for hibernating insects and egg masses.

Though autumn's pace seems less urgent than that of spring, birds still pass through in waves. Countless blue jays flow past the Holiday Beach hawk tower in October, for example, where more than 400,000 migrate through each fall. Experienced birders keep tabs on wind direction and cold fronts to predict major flights. Whereas spring migrants choose rising

If mud flats are available for foraging, short-billed dowitchers can sometimes be seen by the hundreds. They are common late summer/early fall migrants and are usually juveniles like this one. Smaller numbers of long-billed dowitchers come through from late August to November after their shorter-billed cousins have moved south. The two were once considered a single species and still present an identification challenge for birders.

American golden-plovers are long-distance champions among the shorebirds. Many adults fly nonstop over the western Atlantic Ocean to their wintering grounds in South America. Juveniles like this bird, as well as a trickle of adults, travel south through mid-America. Golden plovers are much more common in spring than in fall, using recently plowed fields as sources of food.

temperatures and south winds found on the western sides of high-pressure systems, fall travelers begin to move when the wind is from the north or northwest and the temperature is dropping after a cold front has passed. They tend to avoid rain, fog, high winds, and disturbed weather in general. To arrive safely without straying off course and without burning too much precious fuel, birds must work prudently with, rather than against, wind and weather. Of songbirds, 99 percent have left Ontario before September 1, with most gone by mid-August.

Northern harriers lead the ranks of migrating raptors. They may appear as early as mid-August, well ahead of most other birds of prey. However, by the end of the month the first ospreys, broad-winged hawks, and American kestrels wing past observation points at Hawk Cliff and Holiday Beach, Ontario, on the northern shore of Lake Erie. Lake Erie Metropark, eight miles from Holiday Beach as the raptor flies across the mouth of the Detroit River in Michigan, is another prime hawk-watching location.

All three are excellent places for viewing migrating hawks and learning to identify them in flight. The reason for the large numbers is geographical and relates to some raptors' reluctance to fly over water. Harriers, ospreys, merlins, and peregrine falcons are strong fliers with tapering wings that allow them to strike out boldly across large bodies of water. Observers at Point Pelee often see them take off from the tip of the point and fly directly across the lake. In good weather, even a few American kestrels and sharp-shinned hawks will do the same.

But soaring raptors, such as turkey vultures and hawks in the buteo group(broad-winged, red-shouldered, and red-tailed), almost never cross water. Their shorter, broader wings are ideal for exploiting columns of warm, rising air called thermals as a way of covering distance. A turkey vulture can soar without flapping because it circles within a thermal, always gliding downward but being carried upward at the same time by the more rapidly rising air bubble. Naturalist Edwin Way Teale compared thermal soaring with the lift that occurs when a person walks slowly down a rapidly rising escalator. When the bird finally reaches the top of the air column, it partially folds its wings and glides to the base of another thermal in the direction of its travel, where it repeats the process. Traveling in numbers helps vultures and hawks locate thermals by watching for other birds spiraling ahead of them

Since water is cooler than land in both spring and fall, thermals do not form over the lake, and soaring raptors must detour around it. A map will show that the peninsula of southwestern Ontario funnels these birds southwest along Erie's north shore and around its western end, creating spectacular concentrations at certain points in the fall. (Hawks can also be seen migrating in spring, but not in such large numbers because of differences in their spring routes north.)

Raptor experts have recognized Lake Erie Metropark's importance as a hawk observation site for only about fifteen years, yet watchers at the site consistently tally more raptors than those at either Hawk Cliff or Holiday Beach. Although there is

Common snipe winter throughout South America and are common migrants here, though they are more often seen in spring than in fall with a few remaining to nest. Like their relative the American woodcock, they wear cryptic colors that help them to escape the notice of predators. These include humans: Snipe and woodcock are the only two shorebirds that are still classified as game birds and may be legally shot. The two species are also alike in their sensitive, probing beaks and in their remarkable fields of vision, which cover nearly 360 degrees.

This adult Lapland longspur has exchanged its brighter black, white, and chestnut breeding plumage for more subtle coloring. Summering in the circumpolar high Arctic tundra, these birds arrive here in October and flock with snow buntings and horned larks in winter fields. As with snow buntings, their thick, fluffy feathers help them withstand the season's cold.

An adult black-bellied plover forages at Ashtabula Harbor, Ohio. It is well into molt from summer plumage to its basic winter coat of sober gray. The transition between the two plumages proceeds relatively quickly. The black-bellied is slightly larger than its close relative, the American golden-plover, whose back feathers are more gold than gray, and it is a common migrant in the Great Lakes area.

Semipalmated sandpipers are perhaps North America's most common shorebirds, sometimes passing through our region by the thousands. The species is one of five small sandpipers called "peeps": semipalmated, western, least, white-rumped, and Baird's sandpipers. Difficult to distinguish from each other, all five migrate through the area. This individual is a juvenile and, along with others of its species, will leave the area by the end of September. Pointe Mouillee in Michigan is one of a handful of locations hosting numerous shorebird migrants.

some disagreement among birders, this is probably because hawks following Erie's north shore and other birds flying south along Lake Huron's western coast all converge here just south of Detroit. (Counters probably miss many birds, which thermals carry out over the lake or farther inland.) In addition, observers at the metropark are treated to birds flying much lower than at Holiday Beach because they have lost their lift over the short water crossing of the Detroit River.

Whatever the reasons, a peak day at the metropark can be extraordinary, especially at the height of the broad-winged hawk migration. Broad-wingeds, along with their western cousins, Swainson's hawks, are famous long-distance travelers among the raptors. The broad-wingeds that soar above Lake Erie Metropark are en route to a wide flyway extending along the Mississippi River, the Texas coast, and eventually into Mexico, Central America, and northern South America. They arrive in our area between September 10 and 20, usually around the fifteenth.

The birds' timing must be impeccable. Ornithologists believe that these hawks, which generally hunt from stationary perches, seldom feed during migration, having laid down fat to tide them over until they reach the wintering grounds. To arrive there, they must conserve this stored energy by soaring in thermals, which they often do in great flocks called kettles. Some kettles can hold thousands of birds. If the broad-wingeds wait until too late in the year, thermals will weaken, and the exertions of flapping flight may fatally deplete their reserves. As ornithologist Keith Bildstein puts it, broad-wingeds are in fact, "racing with the sun." His account of the constraints of broad-winged migration in Kenneth Able's *Gatherings of Angels* is excellent reading for those who want to understand this remarkable natural event.

In the Lake Erie region, broad-wingeds afford the most dramatic moment of the fall migration, and for lucky or foresighted birders it can be very dramatic indeed, depending on weather conditions. Before 1999, the highest number of broad-wingeds reported at Lake Erie Metropark in a single season was 399,000, hardly a number to scoff at! However, in the course of one spectacular day, September 17, 1999, birders at the park estimated an unbelievable 503,000 broad-wingeds soaring high past nearby Pointe Mouillee around the western end of the lake on the way to their South American wintering grounds. Of such

The feather wear on this semipalmated plover (not to be confused with the sandpiper of the same name) indicates that it is an adult bird. Unlike its cousin the piping plover, which is almost extirpated in the Great Lakes area because of declines in its beach nesting habitat and disturbance by people, the semipalmated is a fairly common migrant. The location of its nesting grounds in far northern Canada and Alaska has spared it the disturbances that have plagued the piping plover.

fortuitous occurrences legends are made! By the end of the month, however, all the broad-wingeds except stragglers have passed on to the south.

Besides broad-wingeds, harriers, and kestrels, accipiters such as sharp-shinned and Cooper's hawks appear in September. Red-tailed and red-shouldered hawks soar by in the weaker thermals of that month, and merlins and a few peregrine falcons also arrive. Most species do not span such great distances as broad-wingeds, whose goal is South America, though some kestrels and Cooper's hawks may move as far south as the Gulf States. Most stay farther north. Some peregrines, however, travel from the Canadian Maritimes to Patagonia—in a little over a week. Banded "sharpies," as birders like to call sharp-shinned hawks, have been recovered in Costa Rica, Guatemala, and Mexico. Unlike broad-wingeds, these bird-catching little hawks hunt as they migrate, as do the slightly larger Cooper's hawks and the falcons.

Some red-tailed hawks arrive as late as November, and many may spend the winter in Ontario or move east through Ohio to Pennsylvania and New York after rounding the lake. Like rough-legged hawks—big buteos from the far north that also enter the Lake Erie region in November—red-tails can winter here where snow is usually not so thick as to hide the runs of voles and other small mammals.

In late October and November, numbers of hawks are fewer than at the broad-wingeds' peak in September, but the species variety is greater. Now red-shouldered hawks and turkey vultures make their big push. It's a good time to start looking for golden eagles, as well. Birders have tallied over 240 at Lake Erie Metropark in an exceptional year. Less experienced hawk watchers can hone their identification skills as the experts count

Big roosts of American crows are often seen around Lake Erie in winter. These birds have adapted to suburbia, their large roosting numbers sometimes creating Jackson Pollock paintings on sidewalks and driveways. In Essex County, Ontario, roosts may grow into the tens of thousands. Humans' garbage and road kill are much to the liking of these intelligent and omnivorous birds.

and identify the birds passing over, gathering data important for assessing the status of raptors in North America: "Harrier at 11:00 and three TV's coming over the house with the peaked roof." "Kettle of broad-wingeds to the left." "Male kestrel straight up." "'Tail just above the trees." Depending on the winds, the birds may fly either higher or lower, providing both a thrill and a chance to distinguish finally the difference between a flying "sharpie" and a "Coop."

Though most insectivorous birds have passed through by October, common redpolls may appear irregularly, as do other seed-eating birds like evening grosbeaks and pine siskins (also irregularly), fox sparrows, and Lapland longspurs. Eastern bluebirds, which have shifted from eating insects to their winter diet of berries, arrive, and snow buntings make their appearance late in the month.

October and November are the months for gulls and waterfowl. Canvasbacks, both species of scaup, common goldeneyes, and buffleheads migrate eastward through the Niagara River Gorge between Lake Erie and Lake Ontario. Large flocks of tundra swans wing diagonally south across Erie bound for the windy Chesapeake Bay. Loons also migrate, but on a wider front, sometimes gathering in large rafts on local reservoirs. The western end of Lake Erie seems to be a pivotal point for many loons coming down from the north. By the end of November, tens of thousands of red-breasted mergansers can be seen staging along the Lake Erie shore.

Wintering gulls begin to migrate westward into Erie's western basin. This is a good time to look for rarer species like Iceland, glaucous, and Thayer's gulls and the occasional jaeger, a predatory relative of the gulls. Treks to the Niagara River Gorge can be quite rewarding at this time of the year. Dubbed by local birders as the continent's gull capital, this location can offer fourteen species on a good weekend; however, one must patiently sort through the thousands of ring-billed, herring and Bonaparte's gulls to find their rarer cousins. Other specialties of the gorge like red-throated loons, red-necked grebes, and sea ducks such as scoters and long-tailed ducks are also on birders' wish lists now.

Owls, too, are on the move: northern saw-whets in October, long-eared owls in November, and during December short-eared

(*Opposite below*) Dunlins make up a large percentage of shorebirds seen here each year. These round-shouldered little birds are among the most common migrant shorebirds in the region and some of the last to retreat south, with a few usually lingering into December.

Snow buntings are open-ground feeders with thick, insulating feather jackets. They nest in summer on Arctic shores, tundra, rock cliffs, and scree slopes. If the winter is severe, flocks of thousands appear along with horned larks and Lapland longspurs. This fluffed-up bird is a female; she is wearing buffier nonbreeding plumage than her predominantly black and white summer garb.

owls and a few erratic snowy owls in years when lemmings are scarce in the north. The last red-tailed hawks, waves of Bonaparte's gulls, and hordes of blackbirds pass in November. Over the years, the Point Pelee Visitors' Center reports many rarities as well during that month: northern gannet, harlequin duck, purple sandpiper, red phalarope, black-legged kittiwake, ash-throated flycatcher, Townsend's solitaire, western tanager, and mountain bluebird to name a few. December hosts irruptions of winter finches such as common redpolls, evening grosbeaks, and crossbills during certain years. These irruptions are discussed in the next chapter on birds in winter. Lake Erie's greatest effect is felt on the Erie islands. The lake moderates autumn temperatures, which causes late insect hatches. These hold back autumn migrants, especially hermit thrushes, yellow-rumped warblers, and ruby and golden-crowned kinglets.

Most people, even those who know little about birds, are aware of the spring migration, an event that can wake up the most oblivious of us. Fall migration is a much more subtle and

A young Caspian tern begs food from its parent. Identified by a thick, blood-red bill and large size, this species is a common migrant here in both spring and fall. Birders often hear the harsh call before they spot the bird itself. Caspian terns have usually left our region by the end of October.

serendipitous affair as birds quietly and modestly progress toward their southern homes. It's often hard to predict where one will see numbers of fall migrants: 600 loons may appear in one spot, but only one or two in another. One observer may thrill to a thousand tundra swans, while another watcher may find none. Allen Chartier of the Holiday Beach Migration Observatory has likened Lake Erie's unique geographic and climatic effects to a pinball machine, with winds bouncing flocks of birds hither and yon as they pass through the region, flocks appearing suddenly and as suddenly gone.

Yet autumn has its own rewards for birders who like to challenge themselves to identify birds in nonbreeding plumage, who travel to witness the drama of hawk migration, who are attuned to the wild calls and subtle plumage of the "wind birds," as Peter Matthiessen called shorebirds, or who just want to hear the lonely calls of a flock of migrating tundra swans high above the clouds on a snowy day.

ADDITIONAL READING

Able, Kenneth P., ed. *Gatherings of Angels: Migrating Birds and Their Ecology*. Ithaca: Cornell Univ. Press, 1999.

Chartier, Allen and David Stimac. *The Hawks of Holiday Beach*. Self-published, 1993.

Dunne, Pete, David Sibley, and Clay Sutton. *Hawks in Flight*. Boston: Houghton Mifflin, 1988.

Elphick, Jonathan, ed. *The Atlas of Bird Migration: Tracing the Great Journeys of the World's Birds*. New York: Random House, 1995.

Matthiessen, Peter. *The Wind Birds: Shorebirds of North America*. Shelburne, Vt.: Chapters, 1994.

Powers, Tom. *Great Birding on the Great Lakes*. Flint, Mich.: Walloon, 1999.

Snyder, Noel and Helen Snyder. *Raptors: North American Birds of Prey*. Stillwater, Minn.: Voyageur, 1991.

Theberge, John B., ed. *Legacy: The Natural History of Ontario*. Toronto: McClelland and Stewart, 1989.

Thurston, Harry. *The World of the Shorebirds*. San Francisco: Sierra Club, 1996.

Against the Odds

The north wind doth blow,
And we shall have snow,
And what will poor robin do then,
Poor thing?
He'll sit in a barn,
To keep himself warm,
And hide his head under his wing,
Poor thing!

—Anon

How do birds survive when the wind shrieks and snow piles into drifts? How do they cope with the hard season whose very name evokes bleakness and deprivation? Many insects, mammals, reptiles, and amphibians sleep away the cold months in burrows or pond mud. Birds, however, have no such option. They must either flee winter or face it in the open fields, woods, or on the lee sides of hills and buildings.

Human conceptions of winter bird life swing between Anon's rather sentimental view of "poor robin" and the notion that birds are used to getting along in the winter, that they're adapted to it. Reality lies somewhere in between. Many birds are, in fact, uniquely suited for winter survival, with sophisticated circulation systems and natural down jackets fluffed by tiny muscles at the feather roots. (Most birds have 20 to 30 percent more feathers in winter than in summer.) Northern birds like snow buntings and Lapland longspurs are especially well insulated. Yet winter conditions, especially extremes of cold and snow, often stress birds beyond their limits and kill large numbers of them.

Good numbers of white-winged crossbills occasionally appear in the region. These erratic winter invasions may coincide with poor northern cone crops. Winter crossbill flocks include many juveniles and females, as well as a few splendidly colored males like this bird.

(*Opposite*) Severe cold snaps can bring large numbers of winter irruptives into the region like this spectacular flock of common redpolls. When weather improves, these flocks will quickly disperse.

Golden-clad evening grossbeaks appear at feeding stations in small numbers every winter. During flight years, hundreds may be seen, a treat for winter birders. The years 1993–94 produced the last notable invasion, with good numbers reaching as far as southern Ohio.

Evolution has given birds many ways to cope with winter, and their strategies add greatly to the interest of winter birding for observers equipped with thermal underwear and a strong sense of mission. Ruffed grouse, for example, burrow into snowbanks, where, at a depth of two feet, body heat can raise the temperature of their chambers as much as sixty degrees above that of the surface. However, cold alone is not the main problem facing birds, which, after all, developed thermal underwear's prototype, enhanced by natural waterproofing. Starvation is their true winter enemy at a time when food supplies diminish or are covered with snow, daylight hours for foraging are short, and birds' bodies must use more energy to keep warm.

Birds' two extremes of dealing with the cold, of course, are migrating south on the one hand and sticking to the home range throughout the cold season on the other. Most insect eaters must migrate, especially those like swallows, nighthawks, and swifts that hawk for airborne insects. Warblers, vireos, and others, also dependent on insect meals, disappear as well. Only gleaners like woodpeckers, chickadees, and nuthatches remain, scouring the winter woods for egg masses, cocoons, and hibernating adults.

At the other extreme, some raptors, particularly certain owls, stick close to their nesting territories year-round. A bird that relies on sight, hearing, and lethal quickness must know each cranny and dimple of its territory if it is to kill enough wary rodents to see spring. Winter especially taxes young raptors: Their hunting skills and knowledge of terrain are incomplete, and they must settle for marginal hunting territories because the best are almost always occupied. Estimates show that as few as 10 percent of young red-tailed hawks live long enough to don the adults' rusty tail feathers.

But these extremes present too limited a picture. Flying away south or staying put year-round are not the only possible strategies; many adaptations lie between. Fishing birds like kingfishers and great blue herons may tarry around Lake Erie throughout the winter, especially its western end, if weather is mild and

Ground blizzard conditions are especially hard on open-country birds like horned larks, and many perish because food sources are covered with ice. Horned larks are frequent visitors to bird feeders during these hard times.

A screech owl sits in its winter roost tree in an Ohio cemetery, taking in the morning sun. Cemeteries provide important green spaces for breeding birds in an increasingly urban environment.

This short-eared owl rests in a field at Maumee Bay State Park in Ohio. Groups of ten or more birds may gather in rodent-infested areas. The best time to see these birds is just before sunset.

(*Opposite*) A red-tailed hawk surveys its hunting territory. Younger birds are forced into less productive areas and, during severe winters, many perish.

streams and lakes remain unfrozen. During harsher winters, many of these birds drift southward only until they reach open water for fishing. Some individuals, on the other hand, may travel far south of the border into Mexico and Central America.

Many other species arrive in the Lake Erie area from northern summering grounds and replace birds that have migrated south. Red-tailed and rough-legged hawks fly until they reach land sparsely covered with snow. There rodents cannot skulk invisibly

Red-bellied woodpeckers are birds of the southern forests that have pushed their range north during the last century. Their loud calls and tappings are common sounds in wooded areas on winter mornings, especially on the south shore of Lake Erie.

Birders see another winter visitor, the northern shrike, in small numbers every winter. Greater interest in birding has raised the frequency of sightings in recent years.

beneath the white cover and must cross open spaces where hawks can better snatch them up. Red-tails are common here in winter, and smaller numbers of American kestrels and rough-legged hawks join them, especially when heavy snow covers wintering grounds farther north. Cooper's hawks have rebounded from the years of DDT poisoning and now commonly visit birdfeeders, where starlings and mourning doves seem to be their favorite prey.

Raptors are not the only northern birds that move into our area in winter rather than away from it. Northern seed-eaters also sojourn here in numbers. Snow buntings, common on the arctic tundra in summer, drift like snowflakes over our winter fields, whistling thinly and flashing their striking white wing patches. A few Lapland longspurs often accompany these flocks. Large numbers of American tree sparrows, often called "winter chippies," also arrive from northern Canada, fluttering in thickets, hedgerows, and weedy fields. Small feeding flocks of dark-eyed juncos flirt their white tail feathers on snowy field margins and woodland edges. In this species, the slightly larger males winter farther north than the females—their body size gives them a small but important survival edge in cold weather.

Many tree sparrows winter in this region, especially in river valleys south of the lake. Old fields and swampy areas are good places to see these winter residents.

Brown creepers are some of the last birds to leave in autumn and some of the first to return in spring. Small numbers can be found in the feeding flocks during mild winters.

Some blue jays, like this handsome, raucous bird, remain in the region all winter while others migrate farther south. The latter tend to be first-year birds. Blue jays store seeds, acorns, and other foods in tree crevices and in the ground. Many of their stashes are filched by hungry squirrels.

A common feeder visitor, the dark-eyed junco is our most abundant wintering sparrow species, at home in open as well as in forested areas. A few pairs breed here in deep ravines, especially on the Ohio side of the lake.

Flocks of subtly colored cedar waxwings appear in widely scattered areas in winter. Waxwings are very social birds, and they may pass a berry back and forth along their line until one bird eats it. Ornamental plantings north of Lake Erie have reduced their winter visits to the lake area.

(*Opposite*) The beautiful little American kestrel is our smallest falcon. Males like this one are more common in the north during winter than are females. These small raptors are commonly seen in suburbs and may even be spotted flying past high-rise office windows in the urban canyons.

If the winter is mild, raspberry-washed male purple finches and their heavily streaked brown mates may appear at feeding stations. During severe seasons, this species winters farther south. Like all winter finches, they are quite mobile, often lake-hopping when snow is heavy on Erie's north shore.

We often assume that common birds like cardinals, jays, crows, and goldfinches that we see at our winter feeding stations are the same ones we sighted during the summer, but they may well be different birds altogether. Though many cardinals stay in nesting territories all winter, others have been sighted migrating across Lake Erie toward points south. Large flocks of blue jays wing purposefully southwest along the lake (Holiday Beach, Ontario, is a good place to see them), and some goldfinches migrate south, too. Others, in pale, silvery winter plumage devour expensive niger seed all winter at our backyard feeders. Studies have shown that song sparrows from the same brood may either winter along area lakes and rivers or disappear south until we hear their lively, trilling songs again in early March. Many robins may linger all winter as well, especially south of the

Adding touches of bright red to monochromatic winter landscapes, northern cardinals have extended their range well into Ontario during the past hundred years. The more softly hued females sing as well as the males, and walkers can hear the birds' heartening "what-cheer" song as early as January.

White-breasted nuthatches often feed in mixed flocks with chickadees and other small birds. Most seem to stick fairly close to the same range all year, but recent evidence shows that some birds migrate south in the autumn. Nuthatches are the only North American birds that can walk head first down a tree trunk.

lake. The "first robin of spring" may have flocked and roosted with hundreds of others in a protected spot with plentiful berries since the autumn before.

Many other birds also flock together by instinct in winter. Seed eaters such as little brown horned larks scurry over the bleak fields, sometimes accompanied by snow buntings from the Arctic and, to a lesser extent, Lapland longspurs. Tree sparrows and juncos often form nuclei for other mixed flocks in open

Long-eared owls are fairly common residents during the winter, though we seldom see them because of their retiring nature. They are excellent hunters with small rodents being their main fare. These five birds were seen at Ottawa National Wildlife Refuge in Ohio.

country. Busy flocks of chickadees, titmice, nuthatches, brown creepers, downy woodpeckers, and golden-crowned kinglets stake out winter feeding territories in woodlands where they glean the bark for overwintering insects or search out seeds and berries. Each kind of bird forages in its own specialized way, thus reducing competition among species.

Even screech owls may gather in feeding forays of ten or more during very hard winters. Birders can also find well-hidden groups of four or five long-eared owls roosting in thick trees, and assemblages of ten may winter in good hunting areas. Flocking helps young birds survive by imitating their elders' hunting or foraging and makes more efficient use of resources. Many eyes can search for predators, and a small bird in a large flock is statistically safer from such bird-catching raptors as sharp-shinned and Cooper's hawks.

Winter flocks often drift about irregularly, following the vagaries of seasonal food supplies. American crows scavenge opportunistically, and birders commonly note them traveling eastward along Lake Erie's southern shore in winter. Where are they headed, and why? Flocks of highly social cedar waxwings roam about erratically, probably following shifting supplies of seeds and berries that they depend on for survival. It often seems that one either sees no waxwings or a hundred of them. Rosy-red house finches have invaded this region since their illegal release by bird dealers on Long Island in 1940 and have lived here in numbers since the 1970s. These West Coast seed eaters have filled a niche for finches in the East and Midwest, since other winter finches visit here erratically, and they have mostly replaced purple finches as a nesting species. House finches' vivid color and bright, warbling songs liven late winter's drabness, as the birds jockey with house sparrows to pluck seeds from feeders.

Other birds must make more fundamental changes in their seasonal feeding habits. Robins shift from the succulent worms of watered summer lawns to seeds and berries in winter, and late-season storms may even drive them to eat acrid sumac berries as a last resort. Fierce little American kestrels and screech owls shift from a summer diet of insects to the tougher job of pouncing on winter mice and voles. Many young birds that fail to make this difficult change successfully starve before winter breaks. Some birds, like chickadees, nuthatches, jays, and

The downy woodpecker is the most abundant woodpecker species in the region and can live comfortably in many urban settings. It is quite tame and comes readily to suet or peanut butter.

woodpeckers, may begin storing food, such as acorns, in October. How much they recover before the food is stolen or before they forget where it is is unclear.

Even more marginal is the lot of some birds that erupt from northern latitudes to the Lake Erie area during years when food supplies in the north country fail. (Birds "erupt" from the ranges where they nested and "irrupt" into more southern areas.) These unpredictable movements south are not to be confused with migration, which implies a regular round-trip; many irruptive birds never return to their northern ranges.

Severe winters don't necessarily drive these birds to wander south. Again, food shortages usually stress birds more than bad weather does. Snowy owls from the far north appear in the Lake Erie region during years in which Arctic lemming populations undergo their cyclical crashes. Other raptors, such as rough-legged hawks, northern goshawks, and great horned and short-eared owls, also fly southward when food populations cycle downward. Uncommon northern shrikes, also from the Arctic, appear in small numbers around Lake Erie in winter as well. They prey on small birds and mammals and will remain on a good hunting ground until food runs out there. Birders are frequently aware that their own pleasure in seeing unusual birds often comes because the birds themselves are wandering and stressed.

Seed eaters invade the area as well. Finches, especially, respond to failures of cone crops and other seeds by moving south in some autumns. White-winged and red crossbills give birders exquisite glimpses of their rosy coloration, and, in the case of redpolls, ruby-red topknots like small lights winking on and off against the snow as the birds feed. Pine siskins look much like goldfinches and often flock along with them. Pine grosbeaks with rosy feathers and evening grosbeaks with golden bodies and bold white wing patches also appear occasionally at feeding stations south of the lake; however, ornamental plantings in Ontario have reduced their once-memorable numbers farther south. The same is true for Bohemian waxwings. Since each kind of bird specializes in a different food on its northern ranges, invasions of all the winter finches in one year are rare.

Equally grim is the plight of various southern birds during hard winters. These species have extended their ranges north to the Lake Erie region during times of widespread forest clearing, relatively mild winters, and well-stocked suburban feeding stations. Erie lies near the northern edge of the ranges of northern cardinals, tufted titmice, red-bellied woodpeckers, northern mockingbirds, Carolina wrens, and northern bobwhite quail. These birds can thrive during milder winters, but thick snow and late-winter storms may take heavy tolls. The hard winters of 1976–77, 1977–78, and 1978–79 nearly wiped out Carolina wrens across Ohio, and during those years storms suffocated or starved coveys of quail beneath prisoning ice,

causing their numbers to plummet by 90 percent. After occasional severe cold snaps, one can sometimes see dead birds strewn around backyard feeders. They have failed to eat enough to maintain their body temperatures overnight.

Late-season storms also kill birds that can normally survive the bleak trough of the year. These late storms often stoop like a hawk onto the land after warm periods that dry out the berries on which winter robins depend. One late-March storm caught huge flights of Lapland longspurs as they hurtled through Iowa and Minnesota north to their Arctic nesting grounds. At least 750,000 small, snow-covered bodies dotted the ice of two lakes alone, according to author John Kaufmann. Altogether, he writes, millions of longspurs must have perished. March storms may also wipe out early nesters in our area, especially ground-

Another southern species that is gradually extending its range northward is the tufted titmouse. Common along the south shore of Lake Erie, this close relative of the chickadee is only local throughout most of southern Ontario. Its calls, clear and penetrating for such a small bird, are an early sign of spring.

nesting horned larks; ground blizzard conditions are especially hard on these and other open-country birds.

"Poor robin" obviously does more in winter than stay in the barn to keep himself warm. Birds, seemingly so frail, field varied strategies to cope with the wind, snow, and starvation of winter's perilous season. They migrate, or they learn each tiny variation of their hunting grounds. They flock and drift to better their chances of finding food, or they sometimes wander thousands of miles south of their former ranges. Some change from insect to animal or vegetable diets or store food, and many do, like poor robin, seek shelter in thick evergreens or farm buildings when temperatures plummet and winds rise. Hawks sit on their perches for longer periods of time, and many birds move into ravines and other low spots to escape the wind. In such cold weather, foraging burns up too many calories, and staying put becomes a virtue.

Despite these strategies, wind chill often freezes bodies starved for fuel. Snow may drift over early nests, and ice may trap and suffocate the birds themselves. Food may simply run out. So when you see a rosy little house finch gleaming on the tip of a bare branch in early spring, listen respectfully to its passionate warble, and give the tiny bird a small, internal salute.

ADDITIONAL READING

Harrison, Kit, and George Harrison. *The Birds of Winter.* New York: Random House, 1990.

Kaufmann, John. *Wings, Sun, and Stars: The Story of Bird Migration.* New York: W. W. Morrow, 1969.

National Geographic Field Guide to the Birds of North America. 3d ed. Washington, D.C: National Geographic, 1999.

Pasquier, Roger E. *Watching Birds: An Introduction to Ornithology.* Boston: Houghton Mifflin, 1977.

Stokes, Donald W. *A Guide to Bird Behavior.* Vol. 1. Boston: Little, Brown, 1979.

Stokes, Donald W., and Lillian Q. Stokes. *A Guide to Bird Behavior.* Vol. 2. Boston: Little, Brown, 1983.

Appendix

BEST BIRDING SPOTS AROUND LAKE ERIE

The following is a selective list. We decided to include only those places we feel are most significant and most consistently rewarding. Many other locales may be good at certain times or for short periods; these are well scattered around the lake. For more exhaustive information, we suggest consulting these volumes: Clive E. Goodwin, *A Bird-Finding Guide to Ontario* (Toronto: Univ. of Toronto Press, 1995); and Tom Powers, *Great Birding on the Great Lakes* (Flint, Mich.: Walloon, 1999).

LONG POINT WORLD BIOSPHERE RESERVE (ONTARIO)

At 3,250 ha, Long Point offers a unique blend of habitats, including unspoiled beaches, dunes and swales, marshes, ponds, and woodlands. The shallow inner bay was designated a world biosphere reserve in 1986. Though access is restricted, this is perhaps the best site in the Lake Erie region for viewing spring and fall migrations. In spring, thousands of canvasbacks and swans stage offshore at the point. Waves of landbirds arrive in May, and in fall, thousands of migrating waterfowl, gulls, and other water birds pass by the point. Numbers can be phenomenal—200 little gulls in one day! The Long Point Bird Observatory operates three bird banding stations. The area boasts breeding golden-winged warblers.

LAKE ERIE ISLANDS (OHIO & ONTARIO)

Taken as a group, Kelleys, North, Middle, and South Bass Islands and Ontario's Pelee Island are great for migrating land birds, and large numbers of ducks can be viewed around the islands in fall. During that season, warm lake waters create a microclimate that encourages fall migrants to linger. Birders have only recently begun to appreciate properly these stepping stones across the lake. Numerous bird counts are taking place for better coverage of key sites.

NIAGARA FALLS AND RIVER GORGE (ONTARIO & NEW YORK)

Countless thousands of waterfowl, loons, and other waterbirds funnel east through the Niagara River Gorge in autumn. The falls area is the premier gull-watching spot on the continent. On a good day, birders may spot fifteen species, and every year offers its rarities. The gorge is also a fine place for sea ducks, especially long-tailed ducks, and purple sandpipers forage on the rocks along the river. Birding is infinitely better on the Ontario side of the river.

CONNEAUT MARSH (PENNSYLVANIA)

Though accessible only by boat, this high-quality 5,500-acre site is a patchwork of swamp forest, cattail marsh, and open water. Bitterns, rails, black terns, marsh wrens, and bald eagles nest in the secluded setting. Conneaut Marsh, also called Geneva Marsh, is the finest in Pennsylvania.

POINTE MOUILLEE AND LAKE ERIE METROPARK (MICHIGAN)

Pointe Mouillee, a 4,000-acre spit of land lying near the mouth of the Huron River, is one of Michigan's best birding locales. A former shooting club where hunting still goes on in season, Pointe Mouillee is currently experiencing wetland enhancement. Waterbirds and wetland wildlife are major attractions. Many shorebirds visit here if mud flats are exposed. The fall hawk migration here and at nearby Lake Erie Metropark (1,400 acres) offers some of the best hawk watching on this continent. Observers have tallied over 500,000 broad-winged hawks in one day, and golden eagles are regular migrants. In winter, hot water from the Trenton power plant keeps a portion of the waterfront ice free at Lake Erie Metropark. This attracts wintering ducks, geese, swans, and bald eagles. Large rafts of migrating canvasbacks stage here from December through March.

CONNEAUT HARBOR (OHIO)

This choice site in Lake Erie's habitat-starved eastern basin has produced a long list of rarities. Directly adjacent to the marina in downtown Conneaut, the locale is greatly affected by lake water levels. In some years there is little mudflat; in others, the flats can cover many acres. During low-water years, many shorebirds, gulls, and terns appear at close range throughout the summer and fall.

MAGEE MARSH/OTTAWA WILDLIFE AREA COMPLEX (OHIO)

This 6,850-acre complex is a mosaic of open water, marsh, swamp forest, field, and beach. The complex is Ohio's finest birding location. It lies opposite Point Pelee on Lake Erie's south shore, and great numbers of landbirds use it as a stopover in spring and fall. In summer, water levels drop, and shorebirds begin to stop and feed. Ottawa National Wildlife Refuge is one of the few reliable areas in the western basin to see shorebirds. Snowy plover and curlew sandpiper have been observed on the same day at this spot! The spring hawk migration can be excellent. Winter brings hunting hawks and owls, and bald eagles are permanent residents. There is always something to see at this visitor-friendly site.

PRESQUE ISLE STATE PARK (PENNSYLVANIA)

Presque Isle is a 3,200-acre peninsula jutting out into Lake Erie from northwestern Pennsylvania. It preserves beach, sand plain, swamp forest, and clear-water ponds. This scenic area hosts spectacular spring and fall landbird migrations. Fall brings countless flocks of waterfowl, including scoters and long-tailed ducks, migrating past the peninsula. Tundra swans can be seen on the bay—many remain until winter ice forces them south.

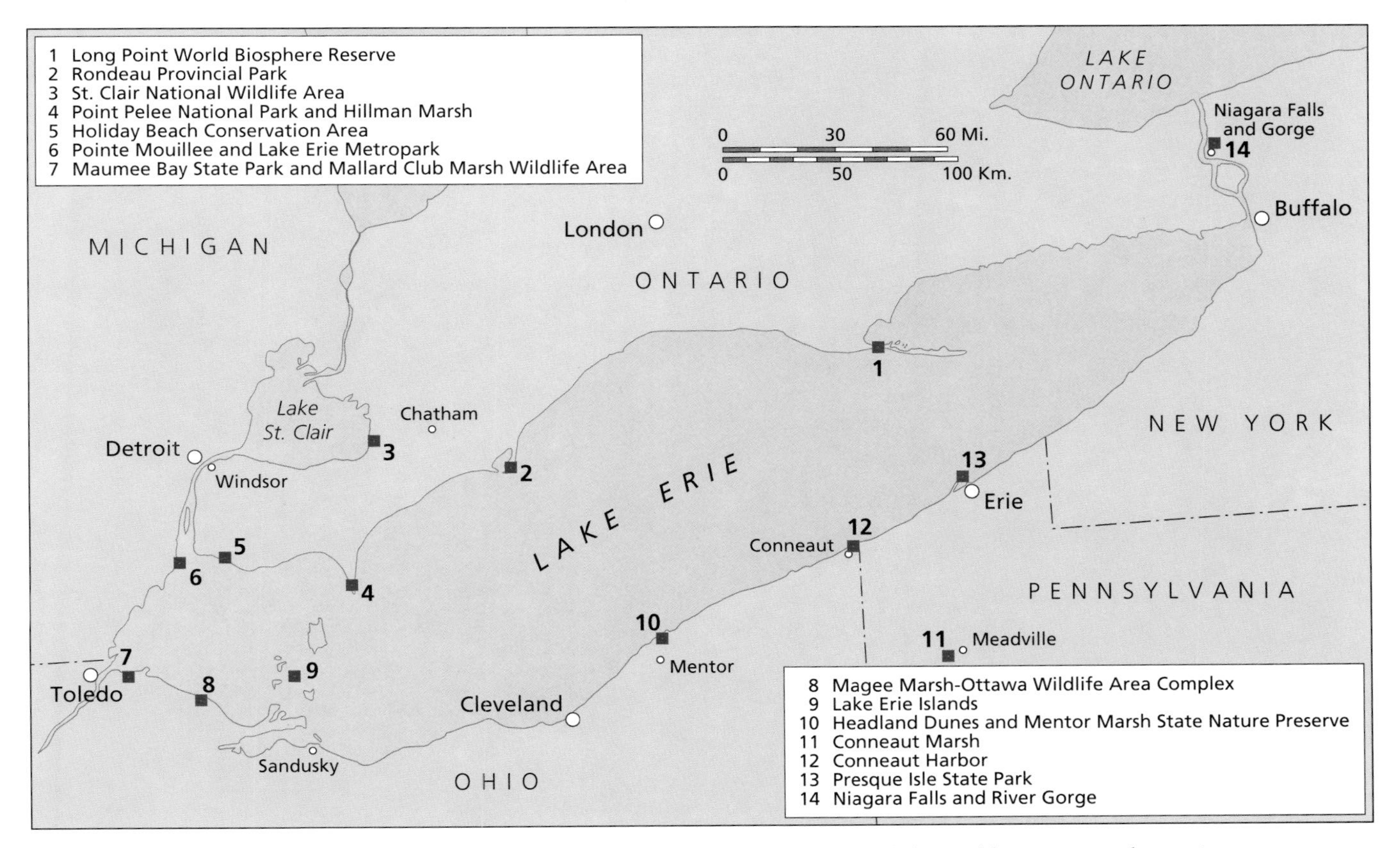

Although Point Pelee, on Lake Erie's north shore in Ontario, is the most celebrated location in the region, many other fine birding spots surround the lake. Because of geography, they tend to be concentrated toward the lake's western end.

HEADLAND DUNES AND MENTOR MARSH STATE NATURE PRESERVE (OHIO)

Mentor Headlands preserves some of the best lake dunes in Ohio. Though the quality of the marsh has declined greatly, the dune area and associated stone jetties have produced an impressive list of over 300 species at this intensively birded spot. From the jetties birders can view gulls, jaegers, and migrating waterfowl in autumn. The landbird migration is also excellent, and many state firsts have occurred here.

MAUMEE BAY STATE PARK AND MALLARD CLUB MARSH WILDLIFE AREA (OHIO)

This 1,529-acre park is a combination of old field, swamp forest, and marsh. Nearby Mallard Club Marsh (410 acres) and Cedar Point National Wildlife Refuge farther east form a contiguous wetland. Maumee Bay's location makes for great spring hawk watching. One can sit outside the visitor center and watch sharp-shinned hawks fly by at eyelevel. The state single-day record (228) was tallied by Gary Meszaros at this site. Winters can be good for hawks and owls. Landbird viewing is best near the visitor center, and the bay can hold thousands of ducks in March. Mallard Club Marsh preserves remnants of cattail marsh,

Lake Erie Metropark in Michigan and Holiday Beach on Erie's northern shore in Ontario are the finest places to watch the tremendous fall raptor migration around the lake's western end. Pictured here is the hawk tower at Holiday Beach.

especially adjacent to Cedar Point National Wildlife Refuge. Bitterns and rails nest at this accessible spot and at Cedar Point (off limits).

ST. CLAIR NATIONAL WILDLIFE AREA (ONTARIO)

St. Clair preserves high-quality cattail marsh. Bitterns, rails, yellow-headed blackbirds, black terns, and marsh wrens are notable nesters in the 289 ha area. King rails are endangered breeders. This is a major staging area for tundra swans, with 10,000 to 20,000 stopping to feed in the adjacent fields every spring.

POINT PELEE NATIONAL PARK AND HILLMAN MARSH (ONTARIO)

Point Pelee, 1,654 ha, together with nearby Hillman Marsh, 362 ha, is an international mecca for birders who converge on the site in spring and rival the numbers of migrating passerines. To be at the "tip" on a fallout day, seeing waves of tanagers, vireos, warblers, and flycatchers working their way through the woods, is memorable indeed. Over the years, birders have reported 355 species in the park, and an experienced observer can easily see over a hundred species in one day. Pelee's sand beaches, forests, marshes, and old fields provide varied habitat, and its tip is a welcome beacon for migrating landbirds. Strong north winds herald the returning migrants with hunting sharp-shinned hawks in close pursuit. The varying habitats of Hillman Marsh, just five kilometers north of Point Pelee National Park, draw migrating waterbirds. Many provincial rarities have occurred at this site.

RONDEAU PROVINCIAL PARK (ONTARIO)

At 3,254 ha, Rondeau is another of Ontario's fine natural areas. Carolinian forests, beaches, dunes, marshes, and open water have created a 300-plus bird list here. The number of spring and fall migrants rivals Point Pelee's, and the list of breeding species is the most extensive in the region. In autumn, large numbers of waterfowl are seen on the bay. Rondeau consistently has the highest Christmas bird count numbers for Ontario.

HOLIDAY BEACH CONSERVATION AREA (ONTARIO)

Holiday Beach hosts some of the best fall hawk watching in North America. Starting in September, thousands of migrating hawks fly by the hawk-watching tower here. The first north winds bring kettles of broad-wingeds, sometimes 50,000 in one day. Peak counts of 1,000 sharp-shinned hawks are not uncommon. October brings buteos, turkey vultures, Cooper's hawks, golden eagles, and an endless stream of migrating blue jays past the 212-ha site. And 520 hummingbirds have been tallied in one day! The fall landbird migration is often the best on the north shore. Banded raptor talks are given at the tower on festival weekends.

Birds of the Lake Erie Region
was designed & composed by Will Underwood in 11/16 Sabon Roman Olstyle and Isadora display type on a Power Macintosh G3 using PageMaker 6.5 at The Kent State University Press; printed by sheet-fed offset lithography in four color process on 157 gsm enamel gloss stock, Smyth sewn and glued into paper covers printed in four color process on 260 gsm C2S artboard finished with gloss film lamination by Everbest Printing Company Ltd. of Hong Kong, China; and published by

THE KENT STATE UNIVERSITY PRESS, KENT, OHIO 44242